Introduction to
Electron and Ion Optics

Introduction to
ELECTRON AND ION OPTICS

POUL DAHL

*Institute of Physics
University of Aarhus
Aarhus, Denmark*

ACADEMIC PRESS New York and London   1973

A Subsidiary of Harcourt Brace Jovanovich, Publishers

COPYRIGHT © 1973, BY ACADEMIC PRESS, INC.
ALL RIGHTS RESERVED.
NO PART OF THIS PUBLICATION MAY BE REPRODUCED OR
TRANSMITTED IN ANY FORM OR BY ANY MEANS, ELECTRONIC
OR MECHANICAL, INCLUDING PHOTOCOPY, RECORDING, OR ANY
INFORMATION STORAGE AND RETRIEVAL SYSTEM, WITHOUT
PERMISSION IN WRITING FROM THE PUBLISHER.

ACADEMIC PRESS, INC.
111 Fifth Avenue, New York, New York 10003

United Kingdom Edition published by
ACADEMIC PRESS, INC. (LONDON) LTD.
24/28 Oval Road, London NW1

Library of Congress Cataloging in Publication Data

Dahl, Poul.
    Introduction to electron and ion optics.

    Bibliography: p.
    1.   Electron optics.   2.   Electron beams.
3.   Particle accelerators.   4.   Ions.   I.   Title.
QC447.D34   537.5'6   72–13611
ISBN 0–12–200650–X

PRINTED IN THE UNITED STATES OF AMERICA

Contents

Preface   *vii*

1. Electron and Ion Optics

1. Introduction   1

2. Geometrical Optics

2. Matrix Formulation   3
3. Rules for Image Formation   6
4. Beam Emittance and the Determinant   11
5. Types and Quality of Image Formation   22
6. Particle-Analyzing Systems   26

3. Scaling Rules and Dispersion Coefficients

7. Scaling Rules   35
8. Formulas for Deflection   38
9. Coefficients of Dispersion   40

4. Fields

10. Field Equations and Paraxial Fields   42
11. Sector, Fringing, and Quadrupole Fields   45

5. Lenses

12.	Electrostatic Lenses	55
13.	Acceleration System	58
14.	Immersion Lenses and Unipotential Lenses	62
15.	Magnetic Lenses	65
16.	Quadrupole and Magnetic Fringing Field Lenses	67

6. Analyzers

17.	Electrostatic Analyzer with Cylindrical Sector Field	73
18.	Sector-Type Analyzer with Electric or Magnetic Field	77
19.	Spherical Analyzer $E(2)$ and Sector Magnets $B(0)$, $B(1/2)$	89
20.	Spectrographs	91
21.	Analyzer with Homogeneous Electric Field	94
22.	Coaxial Cylinder Analyzer	96
23.	Magnetic Lens Spectrometer	101
24.	Orange Spectrometer	102
25.	Sector Magnet with Inclined Boundaries	103

7. Space Charge and Beam Production

26.	Ideal Beam in a Drift Region	110
27.	Focusing of a Nonideal Beam and Various Beam Effects	116
28.	Electron Extraction	120
29.	Ion Extraction	127

Problems 132

Bibliography 142

Index 143

Preface

During the last five years a laboratory course in electron and ion optics has been given to the second-year students of physics at the University of Aarhus, and this book is an extended version of notes providing the theoretical background for this course. It is my hope that the book may serve not only as introductory reading, but that it may also be useful in laboratory design work.

Personally, I have been working with ion-accelerator design and with experiments on atomic collisions, but in writing the book I have tried to outline general principles without emphasizing particular applications.

First-order effects in focusing and particle analysis are treated, and the book includes a chapter on space-charge effects in beam production and focusing. By surveying types of lenses and analyzers, I have tried to illustrate the variety of design possibilities. No attempt has been made to present the various principles and designs in a historical perspective.

At the end of the text some books are listed which may be used for further reading.

I wish to express my gratitude to Charles Macdonald and Torben Rosenquist Sørensen for assistance, careful criticism, and many useful suggestions. I also thank Alice Grandjean and Inge Schmidt for patiently typing the first, second, etc. drafts, and Tove Asmussen for preparing the figures.

My indebtedness to Professor K. O. Nielsen is of a more general nature as he introduced me to the field of isotope separators and ion accelerators by which I aquired a taste for the subject.

1

Electron and Ion Optics

1. Introduction

Electron and ion optics deal with the motion of electrons and ions in macroscopic electric and magnetic fields. The topic is closely related to the development of apparatus for experimental physics and for many other fields of research and technology.

In this book an attempt has been made to establish the general principles on which the design of the optical components used in particle analyzers and accelerators is based. The intention has been to keep within a framework, which is wide enough to accommodate both basic principles and the treatment of components, but which excludes some phenomena and some types of instruments. This framework is defined by the following conditions:

i. The particles are classical particles; i.e., wave properties are not considered. The velocities, however, may be relativistic.

ii. Strong accelerations, which would cause electromagnetic radiation, are not discussed.

iii. The charge and the mass are the only properties of a particle that are taken into account.

iv. The particles move in static fields.

v. A central path in the apparatus lies in an axis or a plane of field symmetry.

vi. Effects of first order are treated. The effects of higher order are discussed, but treated only in a few cases.

vii. The fields considered are paraxial fields in lenses, fields in analyzers, and short fringing fields. For example, the optical properties of the inclined magnetic fringing field are derived and applied to magnet design.

viii. Apart from the last chapter, we shall be treating the low-intensity case in which the interactions between particles in the beam are neglected.

ix. In the last chapter, space-charge effects are treated for ideal beam structures. The treatment here includes the focusing problem, plane, cylindrical, and spherical diodes, and extraction systems with Pierce geometry.

2

Geometrical Optics

2. MATRIX FORMULATION

For the description of paraxial trajectories in a narrow beam we shall use an xyz coordinate system with the z axis following the central path, which may be curved (Fig. 2.1). In the systems to be considered, the fields have symmetries which make it possible to choose x and y directions in such a manner that the projections of a particle motion onto the xz and yz surfaces will describe possible motions of the particle.

A trajectory is fully determined by a geometrical element (x, x', y, y', z) where $x' = dx/dz$ and $y' = dy/dz$, and by the mass, charge, and kinetic energy, i.e., (m, e, T) of the particle.

A group of particles is defined by the values (m, e, T) in a specified, equipotential region. The z axis is a possible trajectory for particles in the central group (m_0, e_0, T_0).

The image formation, which shall be outlined in Section 3, is performed by trajectories for particles in the central group.

When a particle does not belong to the central group, its trajectory starting in $(x_1, x_1', y_1, y_1', z_1)$ may deviate from the trajectory starting

FIGURE 2.1

2. GEOMETRICAL OPTICS

in the same element for a particle from the central group. This deviation is called dispersion. Normally, the dispersion is evaluated as the deviation from the z axis of a trajectory starting in $(0, 0, 0, 0, z_1)$.

The dispersion is called energy dispersion when

$$(m, e, T) = (m_0, e_0, T_0 + \Delta T),$$

mass dispersion when $(m, e, T) = (m_0 + \Delta m, e_0, T_0)$, and charge dispersion when $(m, e, T) = (m_0, e_0 + \Delta e, T_0)$. The relationship between the three kinds of dispersion is treated in Section 9, while for the present we shall assume $(m, e) = (m_0, e_0)$ and include only the energy dispersion.

In the motion of a particle through a field region with potential $V(x, y, z)$, the total energy $T + eV$ is conserved, and thus the kinetic energy is a function of position $T(x, y, z)$. In the following, however, we shall be dealing with the "axis energy" $T(z)$ which is

$$T(0, 0, z) = T(x, y, z) + e[V(x, y, z) - V(0, 0, z)].$$

The axis energy at each z is the same for all particles belonging to one group, and the group may be identified by $\gamma(z) = [T(z) - T_0(z)]/T_0(z)$, where $T_0(z)$ is the axis energy for the central group. Since $T(z) - T_0(z)$ is independent of z, it is seen that $\gamma(z) T_0(z)$ has a constant value.

The central group has $\gamma(z) = 0$, and for a noncentral group, the dependence of γ on z is determined by the axis potential $V_a(z)$, which is $V(0, 0, z)$. The ratio N between $\gamma_2 = \gamma(z_2)$ and $\gamma_1 = \gamma(z_1)$ is given by

$$N = \frac{\gamma_2}{\gamma_1} = \frac{T_0(z_1)}{T_0(z_2)} = \frac{T_0(z_1)}{T_0(z_1) + e[V_a(z_1) - V_a(z_2)]} \qquad (2.1)$$

The method of using matrix calculation for paraxial trajectories will now be described. A trajectory is fully determined by the set of values $(x_1, x_1', y_1, y_1', \gamma_1)$ at z_1. Then x_2, etc., at z_2 can be expressed as Taylor expansions in these values. A paraxial trajectory has small values of x and y throughout the apparatus, and therefore $x_1, x_1', y_1, y_1', \gamma_1$ are all small, which means that only first-order terms in the series expansions for x_2, etc., are significant. The relation between $(x_2, x_2', y_2, y_2', \gamma_2)$ and $(x_1, x_1', y_1, y_1', \gamma_1)$ is then given by five linear equations, and the properties of the system between z_1 and z_2 may be expressed by a matrix of rank 5. One of the equations is seen to be

$$\gamma_2 = 0x_1 + 0x_1' + 0y_1 + 0y_1' + N\gamma_1.$$

2. MATRIX FORMULATION

The field symmetry allows the motion to be resolved into the two projected motions. The matrix equation for each projection has the rank 3.

The equation for the yz surface is

$$\begin{bmatrix} y_2 \\ y_2' \\ \gamma_2 \end{bmatrix} = \begin{bmatrix} A & B & K \\ C & D & L \\ 0 & 0 & N \end{bmatrix} \begin{bmatrix} y_1 \\ y_1' \\ \gamma_1 \end{bmatrix}. \tag{2.2}$$

The rules for image formation are derived in Section 3 from the equation

$$\begin{bmatrix} y_2 \\ y_2' \end{bmatrix} = \begin{bmatrix} A & B \\ C & D \end{bmatrix} \begin{bmatrix} y_1 \\ y_1' \end{bmatrix}, \tag{2.3}$$

and the dispersion is expressed by

$$\begin{bmatrix} y_2 \\ y_2' \end{bmatrix} = \begin{bmatrix} K\gamma_1 \\ L\gamma_1 \end{bmatrix} \tag{2.4}$$

since (y_1, y_1', γ_1) is normally taken to be $(0, 0, \gamma_1)$.

Consider now a system consisting of two subsystems, one going from z_1 to z_2 and the other from z_2 to z_3. We shall apply an abbreviated notation, in which the matrices of the two subsystems are written as $[z_2 \leftarrow z_1]$ and $[z_3 \leftarrow z_2]$, respectively, and similarly the matrix of the total system is written as $[z_3 \leftarrow z_1]$. One then has the relationship

$$\begin{bmatrix} y_3 \\ y_3' \\ \gamma_3 \end{bmatrix} = [z_3 \leftarrow z_2] \begin{bmatrix} y_2 \\ y_2' \\ \gamma_2 \end{bmatrix} = [z_3 \leftarrow z_2][z_2 \leftarrow z_1] \begin{bmatrix} y_1 \\ y_1' \\ \gamma_1 \end{bmatrix},$$

which shows that the matrix of the total system is given by the matrix equation

$$[z_3 \leftarrow z_1] = [z_3 \leftarrow z_2][z_2 \leftarrow z_1]. \tag{2.5}$$

Note that the commutation rule is not valid for matrix multiplication. The order of subsystem matrices corresponds to the order of subsystems in a figure with the z axis oriented to the left (as in Fig. 2.1).

Frequently, a field-free region occurs as subsystem. If the system between z_1 and z_2 is such a region, one has

$$[z_2 \leftarrow z_1] = \begin{bmatrix} 1 & z_2-z_1 & 0 \\ 0 & 1 & 0 \\ 0 & 0 & 1 \end{bmatrix}.$$

When only the central group of particles is considered, it is sufficient to deal with the imaging matrices of rank 2. Also in this case, the combination rule is given by Eq. (2.5) with $[z_2 \leftarrow z_1]$, etc., now being the imaging matrices.

It will be seen in the following that determinants of imaging matrices occur in certain relations, and from Eq. (2.5) it follows that the determinant of a combined system is the product of the subsystem determinants.

3. Rules for Image Formation

We shall derive the general rules for the image formation, and we assume that the elements in the imaging matrix leading from z_1 to z_2 are known,

$$\begin{bmatrix} y_2 \\ y_2' \end{bmatrix} = \begin{bmatrix} A & B \\ C & D \end{bmatrix} \begin{bmatrix} y_1 \\ y_1' \end{bmatrix}. \tag{3.1}$$

In many cases, z_1 and z_2 lie in field-free regions, although quite frequently it is practical to consider a subsystem with z_1 and z_2 lying inside a field region.

First, we must introduce the concepts of a source space and an image space. In the source space, the rays, which may be virtual, are the tangents to the particle trajectories at $z = z_1$, and we shall use the coordinates x_s, y_s, z_s, where the z_s axis is the tangent to the z axis at $z = z_1$. In this point, we also have $z_s = z_1$. The axes for x_s and y_s coincide with the axes for x and y at $z = z_1$. In the same way, the image space, x_i, y_i, z_i, is introduced at $z = z_2$ (Fig. 3.1).

The extrapolation of rays in the source space (or image space) is performed by means of the matrix of a field-free region.

In the yz projection, let us consider a ray from a source point P at $z_s = z_P$. The starting element of the ray is $(y_s, y_s', z_s) = (y_P, y_P', z_P)$.

3. RULES FOR IMAGE FORMATION

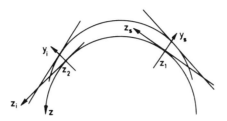

FIGURE 3.1

The ray enters the physical system at $z_s = z_1$, and at z_2, the trajectory is extrapolated as a ray in the image space. The element of the ray at z_i is then given by

$$\begin{bmatrix} y_i \\ y_i' \end{bmatrix} = [z_i \leftarrow z_P] \begin{bmatrix} y_P \\ y_P' \end{bmatrix} \qquad (3.2)$$

where

$$[z_i \leftarrow z_P] = \begin{bmatrix} 1 & z_i - z_2 \\ 0 & 1 \end{bmatrix} \begin{bmatrix} A & B \\ C & D \end{bmatrix} \begin{bmatrix} 1 & z_1 - z_P \\ 0 & 1 \end{bmatrix}. \qquad (3.3)$$

By matrix multiplication we obtain

$$[z_i \leftarrow z_P] = \begin{bmatrix} \alpha(l) & \beta(l) \\ \gamma(l) & \delta(l) \end{bmatrix} = \begin{bmatrix} A + Cl & Al_1 + B + l(Cl_1 + D) \\ C & Cl_1 + D \end{bmatrix} \qquad (3.4)$$

where $l = z_i - z_2$ and $l_1 = z_1 - z_P$. The elements are linear functions of l. It is seen that $\beta(l) = 0$, when $l = l_2$ given by

$$l_2 = -(Al_1 + B)/(Cl_1 + D), \qquad (3.5)$$

and for this l the value of y_i does not depend on y_P', so that all rays from P will pass through one point Q at $z_i = z_Q = z_2 + l_2$. The point Q is the image of P.

Inserting l_2 in Eq. (3.4) we obtain

$$[z_Q \leftarrow z_P] = \begin{bmatrix} d_{12}/(Cl_1 + D) & 0 \\ C & Cl_1 + D \end{bmatrix} \qquad (3.6)$$

where d_{12} is the determinant of the matrix $[z_2 \leftarrow z_1]$

$$d_{12} = AD - BC. \tag{3.7}$$

It should be noted that $\alpha\delta - \beta\gamma = AD - BC$. In general, the matrix for a field-free region does not change the value of the determinant.

In Section 4 it shall be seen that d_{12} may be given in terms of the potentials at z_1 and z_2, when the potential is defined such that one has $T = -eV$.

It is seen that all source points in the plane $z_s = z_P$ have their image points lying in the plane $z_i = z_Q$, and for all points the ratio $-y_Q/y_P$ has the same value, which is the magnification m. It follows from Eqs. (3.6) and (3.5) that m is given by

$$m = -y_Q/y_P = -d_{12}/(Cl_1 + D) = -(Cl_2 + A), \tag{3.8}$$

and the matrix may be written

$$[z_Q \leftarrow z_P] = \begin{bmatrix} -m & 0 \\ C & -d_{12}/m \end{bmatrix}. \tag{3.9}$$

An axial magnification m_z is introduced as $-dl_2/dl_1$, and from Eq. (3.5) we obtain

$$m_z = -dl_2/dl_1 = d_{12}/(Cl_1 + D)^2 = m^2/d_{12}. \tag{3.10}$$

We may now consider a beam from P to Q with the limiting rays (1) and (2) (Fig. 3.2). At P the angle between the two rays is $\theta_P = y_P'(2) - y_P'(1)$, and at Q the angle is given by

$$\theta_Q = y_Q'(2) - y_Q'(1) = (-d_{12}/m)\theta_P.$$

The angle θ_P is the divergence of the beam from P, while $-\theta_Q$ is the convergence of the beam towards Q.

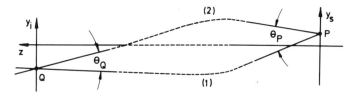

FIGURE 3.2

3. RULES FOR IMAGE FORMATION

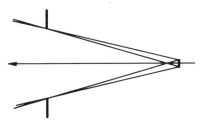

FIGURE 3.3

Thus the transformation of the product $y\theta$ is given by

$$y_Q \theta_Q = d_{12} y_P \theta_P. \qquad (3.11)$$

Let us consider the total beam from an extended source. The beam transmitted to the image is defined by a diaphragm, and the beam shape is the envelope surface for the trajectories. Typically, the source is very small, so that the beam shape is approximately given by trajectories from a central point of the source (Fig. 3.3). In this case Eq. (3.11) gives the relationship between magnification and beam shape.

We shall now introduce the focal points and the principal planes. First, we consider a ray with $y_s' = 0$, ($y_s = $ const). The outgoing ray is $y_i = Ay_s + Cy_s(z_i - z_2)$, and it is seen that $y_i = 0$ when $z_i = F_i$, which is given by

$$F_i = z_2 - A/C, \qquad (3.12)$$

and that $y_i = y_s$ when $z_i = H_i$, which is given by

$$H_i = z_2 + (1-A)/C. \qquad (3.13)$$

The point F_i with $(y_i, z_i) = (0, F_i)$ is the focal point in the image space, and the plane H_i with $z_i = H_i$ is the principal plane. The focal length is defined as $f_i = F_i - H_i$, and it is given by

$$f_i = -1/C. \qquad (3.14)$$

The focal point and the principal plane in the source space are found by setting $y_s = 0$ and $y_s = y_i$ for a ray which, in the image space, is parallel to the axis, $y_i' = 0$. The results are

$$F_s = z_1 + D/C, \qquad (3.15)$$

$$H_s = z_1 + (D - d_{12})/C. \qquad (3.16)$$

The focal length is defined as $f_s = H_s - F_s$. (Note that $f_s > 0$ for $H_s > F_s$, and that $f_i > 0$ for $F_i > H_i$.) It is found that f_s is given by

$$f_s = -d_{12}/C. \tag{3.17}$$

From Eqs. (3.14) and (3.17), the relation

$$f_s/f_i = d_{12} \tag{3.18}$$

is derived.

The elements in the matrix for the system between z_1 and z_2 may be expressed in terms of the focal points and the principal planes, and the matrix is then given by

$$\begin{bmatrix} A & B \\ C & D \end{bmatrix} = \begin{bmatrix} (F_i - z_2)/f_i & f_s - (F_i - z_2)(z_1 - F_s)/f_i \\ -1/f_i & (z_1 - F_s)/f_i \end{bmatrix}. \tag{3.19}$$

When the focal points and the principal planes are known, z_1 and z_2 may be replaced by F_s and F_i, and the matrix $[z_2 \leftarrow z_1]$ by $[F_i \leftarrow F_s]$, which is given by

$$[F_i \leftarrow F_s] = \begin{bmatrix} 0 & f_s \\ -1/f_i & 0 \end{bmatrix}. \tag{3.20}$$

Similarly, one may use the matrix $[H_i \leftarrow H_s]$, which is given by

$$[H_i \leftarrow H_s] = \begin{bmatrix} 1 & 0 \\ -1/f_i & f_s/f_i \end{bmatrix}. \tag{3.21}$$

It is seen that $y_i(H_i) = y_s(H_s) = y_0$. In the special case with $d_{12} = 1$, in which $f_i = f_s = f$, we find $\delta y' = y_i' - y_s' = -y_0/f$. It should be noted here that $\delta y'$ is proportional to y_0 and does not depend on y_s'.

From the properties of the focal points and the principal planes, it is seen that the image point Q may be found by the geometrical construction shown in Fig. 3.4. The construction fails when $z_P = F_s$, but in this case it is seen from Eq. (3.20) that a parallel beam is obtained with $y_i' = -y_P/f_i$.

The figure shows that m is given by

$$m = -y_Q/y_P = q/f_i, \tag{3.22}$$

where $q = z_Q - F_i$.

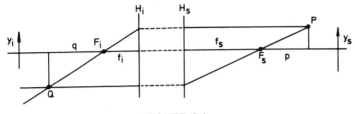

FIGURE 3.4

If, by reversing the rays, we consider Q as the source and P as the image, the magnification n in this image formation is given by

$$n = -y_P/y_Q = p/f_s, \qquad (3.23)$$

where $p = F_s - z_P$.

The product of the two magnifications is unity, $mn = 1$, and thus the relationship between the positions of the source plane and the image plane is given by

$$(q/f_i)(p/f_s) = 1. \qquad (3.24)$$

With the distances measured from the principal planes, $a = H_s - z_P$ and $b = z_Q - H_i$, we obtain the relation

$$f_s/a + f_i/b = 1. \qquad (3.25)$$

Finally, the nodal points are introduced as $(y_s, z_s) = (0, K_s)$ and $(y_i, z_i) = (0, K_i)$ where $K_s = F_s + f_i$ and $K_i = F_i - f_s$. It is easily verified that for the incoming ray $y_s = y_s'(z_s - K_s)$, the outgoing ray will be $y_i = y_i'(z_i - K_i)$, where $y_i' = y_s'$.

The simple relation given by Eq. (3.24) can be recognized as Newton's rule, and it may be mentioned that all relations for paraxial image formation are obviously the same in particle optics and in light optics.

4. Beam Emittance and the Determinant

The structure of a beam of particles from the central group is described at each value of z by a density distribution in the $xx'yy'$ space, where x', y' are defined previously. When the density falls off sharply at a boundary surface, the volume ω_{xy} within this surface is called the beam *emittance*. If x' and y' are small for all trajectories, ω_{xy} is given by

$$\omega_{xy} = \int \Omega \, dx \, dy, \qquad (4.1)$$

where $\Omega(x, y)$ is the solid angle for trajectories through (x, y).

When N particles per unit of time pass through z, the *brightness* (or *luminosity*) of the beam is introduced as N/ω_{xy}.

The emittance may be determined by two beam cross sections, which in the apparatus are defined by means of two diaphragms, or by a source area and one diaphragm. Then, ω_{xy} is the *acceptance* a_{xy} of the optical system.

If ω_{xy} at z_0 lies within the acceptance a_{xy} of the system $z \geqslant z_0$, the whole beam is transmitted through the system.

If diaphragms are inserted in order to define the beam, and a_{xy} at z_0 lies within ω_{xy}, we say that the beam is collimated. Only the fraction a_{xy}/ω_{xy} of the beam is transmitted through the collimator.

Often, the beam is considered in a projection, say the yz projection, and a one-dimensional emittance ω_y is introduced as area in the yy' plane. A beam structure at z is shown in Fig. 4.1, where the dots rep-

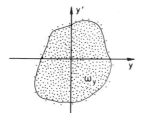

FIGURE 4.1

resent trajectories through the cross section z. If y' is small for all trajectories, the area ω_y is given by

$$\omega_y = \int \theta \, dy, \qquad (4.2)$$

where $\theta(y)$ is the angle for trajectories through y.

Assuming that the whole beam is transmitted, the transformation of ω_y from z_1 to z_2 may be derived from Eq. (2.3).

An element of emittance $d\omega_y(z_1) = dy_1 \, dy_1'$ may be specified by (y_1, y_1'), $(y_1 + dy_1, y_1')$ and $(y_1, y_1' + dy_1')$.

4. BEAM EMITTANCE AND THE DETERMINANT

The vectors $(dy_1, 0)$ and $(0, dy_1')$ are transformed into the vectors $(A\, dy_1, C\, dy_1)$ and $(B\, dy_1', D\, dy_1')$, respectively, and the area $d\omega_y(z_2)$ is the numerical value of the vector product of those vectors. We then obtain

$$d\omega_y(z_2) = \begin{vmatrix} A & B \\ C & D \end{vmatrix} dy_1\, dy_1' = d_{12}\, d\omega_y(z_1). \tag{4.3}$$

The determinant does not depend on y_1 and y_1', and the transformation of ω_y is given by

$$\omega_y(z_2) = d_{12}\omega_y(z_1). \tag{4.4}$$

This includes Eq. (3.11) as a special case.

If $d_{12} = 1$, ω_y is conserved, but the shape of the area varies with z. Let us illustrate this by two examples.

The first example is a drift region:

$$\begin{bmatrix} 1 & z_2 - z_1 \\ 0 & 1 \end{bmatrix}, \quad d_{12} = 1.$$

In Fig. 4.2 the beam is defined by the diaphragms w_1 and w_2 at z_1 and z_2, respectively. We assume, that $z_2 - z_1$ is large compared to w_1 and to w_2. (The figure is not drawn to scale.) From $w_2 = (z_2 - z_1)\theta_1$ and $w_1 = (z_2 - z_1)\theta_2$ we obtain

$$w_2\theta_2 = w_1\theta_1,$$

which is in accordance with Eq. (4.4). The two yy' planes are shown in the figure.

In the second example we consider the transformation from H_s to H_i, Eq. (3.21), when $f_s = f_i = f$;

$$\begin{bmatrix} 1 & 0 \\ -1/f & 1 \end{bmatrix}, \quad d_{12} = 1.$$

It is seen that $y_i(H_i) = y_s(H_s) = y_0$ and $y_i' - y_s' = -y_0/f$. Therefore, w and θ, and thus $\omega_y = w\theta$, are conserved. The shape of ω_y is changed as shown in Fig. 4.3.

14　　　　　　　　　　2. GEOMETRICAL OPTICS

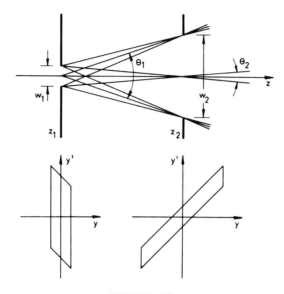

FIGURE 4.2

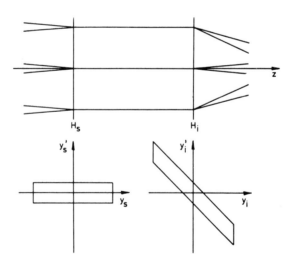

FIGURE 4.3

4. BEAM EMITTANCE AND THE DETERMINANT

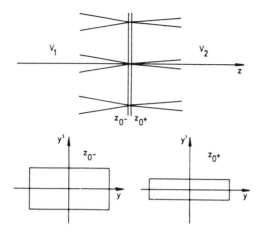

FIGURE 4.4

Consider next the system shown in Fig. 4.4 where an electric double layer produces a jump in potential. The kinetic energy is T_1 for $z < z_0$ and T_2 for $z > z_0$. When a particle passes through the double layer, its transverse components of momentum are not changed; let us assume $p_x = 0$. The trajectory forms the angle θ_1 to the z axis for $z < z_0$ and θ_2 for $z > z_0$. For nonrelativistic velocities we find $p_y = p_1 \sin\theta_1 = p_2 \sin\theta_2$, where $p_1 = (2mT_1)^{1/2}$ and $p_2 = (2mT_2)^{1/2}$. This gives $\sqrt{T_1}\sin\theta_1 = \sqrt{T_2}\sin\theta_2$, which is analogous to the rule for refraction in light optics.

For small beam angles θ_1 and θ_2, we obtain $\theta_1\sqrt{T_1} = \theta_2\sqrt{T_2}$, which together with $w_1 = w_2$ shows that d_{12} in Eq. (4.4) is given by $d_{12} = (T_1/T_2)^{1/2}$. The shapes of the ω_y areas at z_0- and z_0+ are shown in the figure.

In the following we shall introduce a normalized emittance and present the very useful and general relationships one obtains. Here, Eqs. (4.6) and (4.10) will be stated without proof, but they follow directly from the basic discussion, which will be given at the end of the section.

The value of d_{12} is given in terms of the axis potentials $V_1 = V(0, 0, z_1)$ and $V_2 = V(0, 0, z_2)$. Here, the arbitrary constant in the potential is chosen such that the kinetic energy T is given by

$$T = -eV. \tag{4.5}$$

The value of d_{12} is given by

$$d_{12} = (V_1/V_2)^{1/2}, \quad (4.6)$$

which is in agreement with the result for the special system of an electric double layer. Obviously, the same value of the determinant is obtained for the xz projection, and it is seen that $\omega_y \sqrt{V}$ and $\omega_x \sqrt{V}$ are independent of z. By means of Eq. (3.18) we obtain

$$f_s/f_i = (V_1/V_2)^{1/2} \quad (4.7)$$

for both projections.

Instead of plotting y' vs. y as in the above examples, the beam structure at z may be described by plotting p_y vs. y. For the narrow beam one has $p_y = y'p_z \doteq y'p_0$, where p_0 is the momentum of a particle moving in the z axis; this gives

$$p_y \doteq y'p_0 = y'(-2meV)^{1/2}. \quad (4.8)$$

A normalized emittance τ_y is introduced as an area in the yp_y plane. It is related to ω_y through

$$\tau_y = \omega_y(-2meV)^{1/2}. \quad (4.9)$$

Frequently, the adjective normalized is omitted. The emittance τ_y is independent of z. Similarly, τ_x is an area in the xp_x plane, and τ_x is independent of z.

The conservation of τ_x and τ_y is obtained for the particular xyz frame introduced in Section 2. As is recalled, the field symmetry is such that the projections of a moving particle onto the xz and yz surfaces will describe possible particle motions. This is assumed whenever τ_x and τ_y occur in the following.

For the two-dimensional emittance ω_{xy} one has

$$\omega_{xy}(z_2) = d_{12}^2 \omega_{xy}(z_1), \quad (4.10)$$

which shows that $\omega_{xy} V$ is independent of z. The normalized emittance τ_{xy} is the volume in the $xp_x yp_y$ space, and it is independent of z. Normalization of ω_{xy} is performed by

$$\tau_{xy} = \omega_{xy}(-2meV) = \text{const.} \quad (4.11)$$

At each z, the volume τ_{xy} does not depend on the orientation of the xy system in the plane normal to the z axis.

4. BEAM EMITTANCE AND THE DETERMINANT

The ideal beam is a beam in which only particles belonging to the central group are present, and which has $\tau_x = 0$ and $\tau_y = 0$. In principle, such a beam can be produced from a point source emitting particles with a precisely defined energy. By use of suitable optical systems the beam can be projected as a parallel beam ($x' = y' = 0$), even an infinitely narrow parallel beam (the pencil beam), or as a convergent beam giving a point focus.

A source of radioactive material may have a nearly monoenergetic emission, $T_0 = -eV_0$. Consider such a source, which has a width w_0 in the yz projection. Typically, w_0 is small compared to the width of the beam-defining diaphragm; the latter determining the beam angle θ_0. By means of an optical system, the beam is focused in a region where the kinetic energy is $T = -eV$, and the beam angle is θ. We may determine the focus width w by using $\tau_y = w_0 \theta_0 (-2meV_0)^{1/2} = w\theta(-2meV)^{1/2}$, which gives

$$w = w_0 (\theta_0/\theta)(V_0/V)^{1/2}. \tag{4.12}$$

Alternatively, Eq. (4.12) is obtained by inserting Eq. (4.6) in Eq. (3.11).

Another type of source is a cathode releasing electrons with thermal energies. The electrons are accelerated, and in the focus region they have a fairly well defined energy $T = -eV$, which is much larger than the thermal energies. The source emission is then nearly monoenergetic with energy zero.

At the cathode electrons are moving in all directions. There is no well-defined limitation for p_y, since the distribution for p_y has tails going to $+\infty$ and $-\infty$. However, after acceleration the beam is collimated by means of a diaphragm. Here, the brightness is not homogeneous over the area of acceptance in the yp_y plane. Let us assume that a large fraction of the beam is found within some small part of the acceptance; this part ought to be used as beam emittance.

A reasonable range for p_y at the cathode is given by $\delta p_y = 2(-2meV_T)^{1/2}$, where $-eV_T = kT$. Here T is the temperature of the cathode and k is the Boltzmann constant. Using the units volt and degree Kelvin we obtain

$$1 = -eV_T/kT = 1.16 \cdot 10^4 V_T/T, \tag{4.13}$$

which for $T = 1160°K$ gives $V_T = 0.1$ V.

2. GEOMETRICAL OPTICS

With a cathode width of w_0, the included emittance τ_y is given by

$$\tau_y = w_0 2(-2meV_T)^{1/2}, \qquad (4.14)$$

and the corresponding focus width w is

$$w = w_0 (2/\theta)(V_T/V)^{1/2}. \qquad (4.15)$$

The total width may, however, be much larger due to tails of beam brightness. In a high-energy accelerator, such tails must be removed by collimation in a section, where the energy is still low, since if high-energy particles hit a diaphragm, dangerous X-ray production will result.

For obtaining Eq. (4.15) it has been assumed that emittance is conserved even for a beam with high particle density, since the density will be high at least near the cathode. This assumption is verified below where the space-charge model is outlined. Thus, the equation is valid.

It should be recalled, however, that θ is the angle, in the yz projection, for trajectories through a point (y) of the beam cross section at focus, and Eq. (14.5) is a particularly useful relation when θ may be identified with the beam angle obtained without space charge. If we insist on inserting θ as the latter beam angle, Eq. (14.5) is only valid when the particle density is low in the focus region. Then the equation gives the focus width due to source emittance.

In Chapter 7 the space-charge effect is treated for ideal-beam structures with zero emittance; it is also investigated when focus width is mainly due to space charge in the focus region, rather than due to source emittance. Frequently, the space-charge effect is only of importance in a region near the cathode.

The last topic of the survey is motion in the z direction.

The central group has axis energy $T_0 = p_0^2/2m$, and the transverse momentum $p_t = (p_x^2 + p_y^2)^{1/2}$ is for $x = y = 0$ given by $p_t = p_0 \sin\theta$, when the trajectory forms the angle θ to the z axis. For θ small, this gives $p_0 - p_z = p_0(1 - \cos\theta) \ll p_t \ll p_0$. For each (x,y) in the cross section of a narrow beam one has $p - p_z \ll p_t \ll p_0$, and the range δp_z is zero to second order.

With groups in a small range δT around T_0, δp_z is not zero to second order, and one has

$$\delta T = (p_0/m)\,\delta p_z = \text{const} \qquad (4.16)$$

4. BEAM EMITTANCE AND THE DETERMINANT

since δT is independent of z. Thus δp_z must be inversely proportional to p_0.

A short section of beam, δz, passes through a certain z in an interval of time, which is independent of z. This may be seen as follows. Let the section be situated in $(-\delta_0 z, 0)$ at time $t_0 = 0$, and in $(0, \delta_1 z) = (z_1 - \delta_1 z, z_1)$ at the time $t_1 = \delta_0 t$. A second section may at time t_1 be added in $(-\delta_0 z, 0)$, and a third one at time $t_2 = 2\delta_0 t$, and so on. In the interval of time $t_{n+1} - t_n = \delta_0 t$ the first section passes through $z_n = \sum_{i=1}^{n} \delta_i z$.

By means of the relation

$$\delta t = (m/p_0)\,\delta z = \text{const} \tag{4.17}$$

it is found that δz is proportional to p_0.

We have then obtained the interesting result that

$$\delta z\,\delta p_z = \delta t\,\delta T = \text{const.} \tag{4.18}$$

This shows that $\tau_z = \delta z\,\delta p_z$ is independent of z. (The relation may be compared to the uncertainty principle in quantum mechanics, $\Delta z\,\Delta p_z \gtrsim \hbar$, etc.)

In the basic discussion Liouville's theorem will be applied, and let us start by outlining what this theorem states.

Consider a system comprising one particle moving in a static field. The motion is described in Cartesian coordinates, i.e., an ordinary xyz frame is used, not the frame from Section 2 previously used.

The six-dimensional space $xp_x yp_y zp_z$ is called the phase space. The state or phase of the system at a time t_0 is specified by the position and momentum of the particle, i.e., by a point $P(t_0)$ in the phase space. The change in time of the state is described by the curve $P(t)$, which is uniquely determined by $P(t_0)$. This implies that the phase curve $P(t)$ does not intersect any other phase curve.

Next we consider an ensemble of systems, which all have the same static field and one moving particle. Since phase curves do not intersect each other, a laminar streaming of phase points is obtained. The ensemble may include systems with phase points, at time t_0, situated in a small volume $d\tau$ around $P(t_0)$, and the motion of these points was investigated by Liouville. From general principles of particle dynamics, he proved that the volume $d\tau$ of the phase region in invariant, although the shape changes in time. Thus, the motion of phase points is similar to the streaming of an incompressible fluid, and not only the infinitesimal volume $d\tau$, but any phase volume τ is conserved.

If the particle motion is constrained to the x axis, the phase space is two-dimensional, the xp_x plane, and any phase area τ_x in this plane is conserved.

For particle motion in the xy plane, any volume τ_{xy} in $xp_x yp_y$ is conserved. Phase areas τ_x and τ_y are obtained by projecting τ_{xy} onto xp_x and yp_y, respectively. These areas are conserved only for a particular type of system. Assume here the following property: when $(x(t), y(t))$ describe any possible motion, then $(x(t), 0)$ and $(0, y(t))$ will describe possible motions. Evidently τ_x and τ_y are then conserved, and furthermore, one has $\tau_{xy} = \tau_x \tau_y$.

Consider now the narrow beam of particles from the central group. Assuming that mutual interactions are insignificant, the beam can be represented by an ensemble of one-particle systems. For particles from the central group, the phase volume τ is zero, since $p = (2mT)^{1/2}$ is a single-valued function of position (x, y, z).

Let the ensemble be the particles in a short beam section, and let us, for reference, choose one of these, particle A. At each time t we may orient the z axis along the tangent to its trajectory. This gives $\tau_z = 0$ as pointed out in the above survey. The phase points lie in a four-dimensional "surface" normal to the zp_z plane, and the "area" in this surface is recognized as the beam emittance τ_{xy}.

Including groups within dT, the phase volume is bounded by two surfaces separated by $d\tau_z$, which is infinitesimally small, and it is given by $d\tau = d\tau_z \tau_{xy}$. With the z axis referred to particle A, this is valid at each time. In the above survey it was noted that $d\tau_z$ is independent of z, and thus conserved in time. The conservation of $d\tau$ is stated by the theorem of Liouville, and therefore, the emittance τ_{xy} is independent of z.

In Section 2 it was assumed that any particle motion can be resolved as possible motions in the xz and yz surfaces. In this case τ_x and τ_y are conserved, and τ_{xy} is given by $\tau_{xy} = \tau_x \tau_y$.

One can now prove Eqs. (4.6) and (4.10).

Liouville's theorem may be applied for particle position given in arbitrary coordinates, provided certain generalized momenta, related to the coordinates, are used.

4. BEAM EMITTANCE AND THE DETERMINANT

For semipolar coordinates $zr\varphi$, the generalized momenta are $p_z = m\dot{z}$, $p_r = m\dot{r}$, $p_\varphi = mr^2\dot{\varphi}$, where p_φ is recognized as angular momentum.

The transverse motion in an electric field with axial symmetry, i.e., $V(z, r, \varphi)$ independent of φ, Eq. (10.10), may be resolved as motions in r and φ for which $\tau_r = \delta r\, \delta p_r$ and $\tau_\varphi = \delta\varphi\, \delta p_\varphi$ are conserved. It may also be resolved as x and y motions for which τ_x and τ_y are conserved.

Motion in a magnetic field with axial symmetry, Eq. (10.13), is more complicated. It will be seen in Section 15 that the image space is rotated through an angle φ given by Eq. (15.7) relative to the source space, and it may be shown that when this rotation is taken into account, τ_r and τ_φ are conserved, and likewise τ_x and τ_y.

Frequently, circular beam-defining diaphragms are used in a system with axial symmetry. The beam structure is then fully described in one projection, say the yz projection for which the structure at each z is given in the yp_y plane. Alternatively, the rp_r plane may be used.

Liouville's theorem may be applied to an arbitrary many-particle system with arbitrary fields; with N particles phase space will have $6N$ dimensions.

However, when interactions between particles can be taken into account by means of an additional static field, it is sufficient to consider the one-particle system including this field.

In the space-charge model, the force acting upon one particle, due to electrostatic repulsions from all others, is derived from a continuous, static charge distribution, and beam emittance is conserved. The justification of the model lies in the fact that the fast fluctuations of potential energy in the particle system are insignificant for the beam structure averaged over time.

The problem in calculating beam structure is to obtain consistency so that the charge distribution corresponds to the set of trajectories. In Chapter 7 we shall restrict ourselves to ideal structures for which this consistency is easily obtained.

Magnetic interactions between particles occur for very high velocities, and they may be taken into account in an independent-particle model including the magnetic field for a continuous, static current distribution.

Conservation of emittance is obtained even for large beam cross sections, since the phase point for particle A moving in the axis of the apparatus, is well centered in the phase region for the beam. Therefore, trajectories of a beam, for which τ_y is conserved, are not necessarily paraxial in the sense that the linear relationship given by the matrix equation, Eq. (3.1), is obtained. This implies that $\tau_y = 0$ does not necessarily ensure perfect image formation.

Consider an example where $(y_1, y_1') = (y_1, 0)$ gives $y_2 = y_1$ and $y_2' = -y_1/f + by_1^2 + \cdots$. For an incoming parallel beam, paraxial rays are focused at $z_f = z_2 + f$ with $y_f = 0$, but for a nonparaxial ray, one finds $y_f = fby_1^2 + \cdots$, which is called aberration. For each y_f, the slope y_f' is not single valued, but no continuous interval is obtained.

Using, in the example, the angle θ given by the beam shape near z_2 (where the beam is laminar) and the minimum width w of the beam shape, an apparent emittance τ_y^* can be defined as

$$\tau_y^* = w\theta(-2meV)^{1/2}.$$

In the general case with a field region in which trajectories are curved, τ_y^* can be defined at each z by means of rays taken as tangents to trajectories. For the ideal beam with $\tau_y^* = 0$, one has y' proportional to y when the source point is located on the z axis. In a system with aberrations τ_y^* will not be conserved, and normally it increases with z.

If the beam-defining diaphragms permit only paraxial trajectories, one has $\tau_y^* = \tau_y$, and the beam quality is then given by the source emittance.

5. Types and Quality of Image Formation

The image formation for paraxial rays has been treated in one projection in Section 3. Considering now both projections, there are two types of focusing, point focusing and line focusing.

Point focusing is obtained for $z_{Qx} = z_{Qy} = z_Q$, i.e., when the images of P in the two projections have the same z value. The plane $z = z_P$ is imaged with point-to-point correspondence into the plane $z = z_Q$, and the imaging is called stigmatic image formation.

The magnifications m_x and m_y may differ from one another, and, in the special case of $m_x = m_y$, the image is known as the *Gaussian image* for which image points form a pattern geometrically similar to that of source points in the object plane.

A Gaussian image is obtained when the z axis is not curved and the field is symmetric around the axis, i.e., in the axial-symmetry case.

Line focusing is obtained when z_{Qx} differs from z_{Qy}; here we are primarily concerned with one of the projections, say, the yz projection. It is seen that lines in the x direction in the object plane are imaged into lines in the x direction in the image plane $z = z_{Qy}$. A two-dimensional field in which the particles experience no forces in the x direction is an example of a system with line focusing.

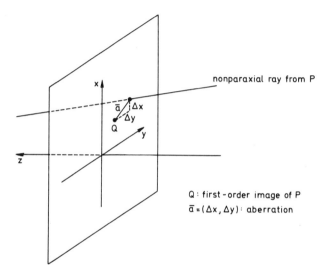

FIGURE 5.1

So far we have dealt with the ideal image formation obtained from the linear theory valid for paraxial rays and shall now consider effects of nonlinearity in the system. A ray leaving the object point P in a particular direction intersects the image plane at a point displaced from the ideal image point Q (or line $y = y_Q$); this displacement is called *aberration* (Fig. 5.1). The quality of the image formation is determined by aberrations for rays in the transmitted beam.

Let us begin by considering a case with P located on the z axis and with a slit diaphragm, permitting only rays with small values of x_P' to be transmitted, while $|y_P'|$ may be large enough for nonlinearity to occur. The aberration Δy may be expressed as

$$\Delta y = by_P'^2 + cy_P'^3 + \cdots, \tag{5.1}$$

and the focusing is called first-order focusing for $b \neq 0$, second-order focusing for $b = 0$, $c \neq 0$, and so on.

Frequently, the approximation $\Delta y \doteq by_P'^2$ may be used for first-order focusing, and it is noted that in this case $\Delta y(-y_P') \doteq \Delta y(y_P')$. For second-order focusing with $\Delta y \doteq cy_P'^3$, it is seen that $\Delta y(-y_P') \doteq -\Delta y(y_P')$. The resulting intensity distributions are illustrated in Fig. 5.2.

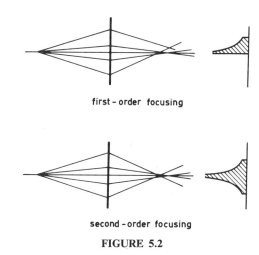

first - order focusing

second - order focusing

FIGURE 5.2

It should be noted that with axial symmetry, the property $\Delta y(-y_P') = -\Delta y(y_P')$ is obtained. This shows that no terms of even order occur in the aberration, and normally, the focusing will be of the second order with aberrations of the third order. This also holds for the additional aberration terms occurring when the object P is not located on the z axis. Effects of some of the third-order aberration will now be briefly described.

With the axial symmetry, the beam from P will normally be defined by a circular aperture, and the yz plane may be chosen such that it contains the object point P. The term $\Delta y \propto y_P^3$ does not destroy the stigmatism, but it causes distortion of the Gaussian image. A square centered on the axis is imaged into a pincushion-shaped pattern, if the

5. TYPES AND QUALITY OF IMAGE FORMATION

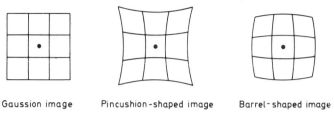

Gaussion image   Pincushion-shaped image   Barrel-shaped image

FIGURE 5.3

coefficient is positive, and into a barrel-shaped pattern if the coefficient is negative (Fig. 5.3).

The terms $\Delta y \propto y_P^2 y_P'$ and $\Delta x \propto y_P^2 x_P'$ will cause astigmatism, i.e., the images in the two projections have different z values. The image point in the yz projection represents a piece of a line in the x direction, and the image point in the xz projection a piece of a line in the y direction (Fig. 5.4). Halfway between the two lines, the cross section of the beam will be circular; this cross section is called the *disc of minimum confusion*. Corresponding to the object plane, three surfaces are defined by the centers of the line images and by the discs of minimum confusion. All three surfaces may be curved.

While distortion is proportional to y_P^3 and astigmatism to y_P^2, the coma effect is proportional to y_P, and Δy includes the terms $y_P y_P'^2$ and $y_P x_P'^2$, while Δx results from $y_P x_P' y_P'$. The effect is rather complex, and we shall only note that it gives an image distribution resembling a comet's tail.

The last geometrical aberration to be mentioned is the spherical aberration or aperture defect due to $\Delta x = c x_P'^3$, $\Delta y = c y_P'^3$. Frequently, the source area is centered on the axis and is so small that it is the only aberration of importance.

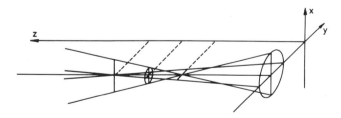

FIGURE 5.4

It is noted that not all the terms have been mentioned, but for an electrostatic field with axial symmetry the coefficients to the remaining terms are all zero. This is not the case in the presence of a magnetic field; the resulting effects, however, shall not be described in this brief survey.

Apart from the geometrical aberrations, the dispersion also causes some smearing of the image when the energy is not sharply defined; this may be referred to as chromatic aberration in obvious analogy to light optics.

Previously, we have defined aberration as the displacement of a single ray from the ideal image point; the result of including all rays from an object point is that an intensity distribution in the image plane is obtained. For an object point located on the z axis, the spherical aberration gives a distribution with an infinite maximum intensity; but inevitably with a finite energy spread, a finite maximum is obtained due to chromatic aberration. A representative width of the image distribution is introduced as the width δy of the interval in which the intensity is higher than half the maximum intensity.

We shall now discuss the possibilities of obtaining information about an unknown source structure from the recorded image. If two object points are brought closer and closer together, it becomes difficult at very small separations to distinguish the two first-order image points. This is easy as long as the separation is larger than $\delta y/m$, i.e., the width δy of the image of a single point converted into a source dimension, while the image must be recorded and examined in great detail for separations smaller than $\delta y/m$. The reciprocal value of this source dimension, i.e., $m/\delta y$, is called the *resolving power* in the image formation.

The solid angle of the beam from an object point is called the *transmission*, and since a large solid angle introduces aberrations, it presents a difficult design problem to obtain at the same time a high transmission and a high resolving power.

6. Particle-Analyzing Systems

A detailed treatment of particle analyzers will be delayed until Chapter 6. In this section, however, the relevant geometrical parameters will be introduced. Furthermore, we shall discuss the choice of the design criteria in an analyzing system.

6. PARTICLE-ANALYZING SYSTEMS

By means of dispersion, a particle-analyzing system separates the particle groups (m, e, T) emitted from a source (Section 2). We shall consider energy analysis for the groups $(m_0, e_0, T_0 + \Delta T)$. The relationship between mass dispersion, charge dispersion, and energy dispersion will be outlined in Section 9.

Now let us consider a deflection system in which the curved z axis lies in a plane containing also the y axis. A point source is placed in $(y, z) = (0, z_P)$. The potential of the image region is equal to the source potential.

For each particle group characterized by T, the image is a line in the x direction, and the image lines for all groups define the focal surface. The intersection of this surface with the yz plane is the focal curve $Q(T)$. An energy scale may be placed on this curve, and the important property of the system is that the different energy groups are separated from one another in the focal surface.

Consider now a small energy range around T_0.

The dispersion from the central group is the distance $Q(T_0)Q(T) = s$, and the coefficient D_s of dispersion is introduced by

$$s = D_s(\Delta T/T_0), \qquad D_s = T_0(ds/dT). \tag{6.1}$$

The coefficient has the dimension of a length.

The dispersion may also be measured in the y direction at $Q(0)$ by using the trajectories with $(y_P, y_P') = (0, 0)$ from the source. The dispersion is then given by

$$y = D(\Delta T/T_0), \qquad D = T_0(\partial y/\partial T). \tag{6.2}$$

In a spectrometer or a "single-channel analyzer" only one detector is used and there is a definite central path from source to detector. The energy T_0 for this path is a function of the strength of the deflecting field. From scaling rules derived in Section 7 it will be seen that for an electric field E one has $T_0 \propto E$, while for a magnetic field B one has $T_0 \propto B^2$. The coefficient D of dispersion will be seen to be independent of T_0. The detector slit defines the interval $|y| \leqslant \frac{1}{2}w$, and the energy interval given by

$$(\Delta T)_{\text{ch}} = (w/D)T_0. \tag{6.3}$$

is called the channel width, and is noted to be proportional to T_0. Such a spectrometer system is treated in the following.

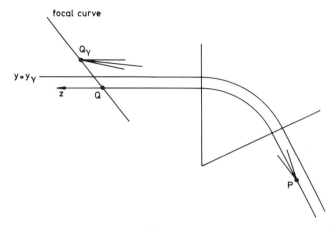

FIGURE 6.1

Dealing with image formation performed by trajectories for $\gamma = \Delta T/T_0$, one of these trajectories must be used as central path. For a broad class of analyzers, this path can be chosen as $y(z) = y_\gamma$ where y_γ is a constant proportional to γ (Fig. 6.1). This will be seen in Section 18 where simple expressions for y_γ are given. The magnification m is to the first order the same as for the central group, and it is immediately seen from the figure that y for the image Q_γ of the source point P on the z axis is given by

$$y(Q_\gamma) = y_\gamma(1+m). \tag{6.4}$$

When m has been calculated, the coefficient of dispersion D is obtained as $\gamma^{-1}y_\gamma(1+m)$.

As to the source, we shall be dealing with two cases: (i) a point source (or a line source) which is so small that the image distribution results from aberrations, and (ii) an extended source which is homogeneous and so large that the first-order image formation may be applied.

Let us consider a simple example with a point source P placed in a homogeneous, magnetic field **B** in the x direction (Fig. 6.2). All trajectories in the yz plane are circles, and the radius r is a known function of the energy $r(T)$. The figure shows the trajectories for the energy T_0. A slit diaphragm determines the transmission angle α, and the direction of a ray from P is given by $y_P' = \tan\theta$, where $|\theta| \leq \tfrac{1}{2}\alpha$. The image Q

6. PARTICLE-ANALYZING SYSTEMS

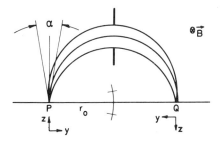

FIGURE 6.2

falls on the diameter (from P) of the circular z axis, and it is seen that y_Q is given by

$$y_Q = 2r_0(1-\cos\theta) = 2r_0(\tfrac{1}{2}\theta^2 - \cdots) \doteq r_0\theta^2 \tag{6.5}$$

for the y axis pointing toward the center.

The image distribution $\rho(y)$ is given by[†]

$$\rho(y) = |d|\theta|/dy| \doteq 1/r_0|\theta| \doteq 1/(r_0 y)^{1/2}, \tag{6.6}$$

and this distribution is shown in Fig. 6.3a.

It is seen that the focal curve is the diameter line from P, and the energy scale is given by $2r(T)$. In the following, we assume that the system is used as a single-channel analyzer with the detector slit w placed at $2r_0$.

Suppose now that particles with a definite energy T_s are emitted from the source. The radius of the trajectories is a function of the field strength $r(B)$, and when B is varied the image distribution moves across the detector slit. The recorded spectrum is the flux through this slit plotted versus $T_0(B)$, which is the energy for which the radius would be r_0. For each B the image position is $D[T_s - T_0(B)]/T_0(B)$. For the monoenergetic source the recorded spectrum is called the *analyzer line shape*. It depends on $\rho(y)$ and w. The interval δT for $T_0(B)$ during which the recorded flux is higher than half the peak value is called *full width at half maximum* and may be denoted FWHM. Note that the relative line width $\delta T/T_s$ does not depend on T_s (see Fig. 6.3b).

[†] Note that $a/bc = a(bc)^{-1}$, while $ab^{-1}c$ may be written as $(a/b)c$ or as ac/b.

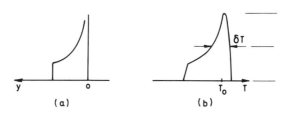

FIGURE 6.3

The *spectral resolving power* is defined by

$$\mathcal{R} = (\delta T/T_s)^{-1}. \tag{6.7}$$

It is easy to establish the existence of two lines T_1 and T_2 in the source spectrum when $|T_1 - T_2| > \delta T$ and difficult when $|T_1 - T_2| < \delta T$.

Let us consider the case with w infinitely small (though this is not a sensible choice for practical applications). The line shape is then the $\rho(y)$ distribution converted into the energy scale by means of Eq. (6.2). The aberration $b = \frac{1}{4}r_0\alpha^2$, Eq. (6.5), corresponds to the energy interval

$$(\Delta T)_b = (b/D)T_0 = (\alpha^2/4)(r_0/D)T_0. \tag{6.8}$$

If the slit is chosen so as to be approximately the same size as the aberration $w \approx b$, the line width δT will be given approximately by $(\Delta T)_b$, and thus the resolving power becomes

$$\mathcal{R} \approx D/b = (4/\alpha^2)D/r_0. \tag{6.9}$$

With the extended source, simple line shapes depending on w may be obtained. The width of the source is Δy_P, and the image width is given by $\Delta y_Q = m\,\Delta y_P$. The line shapes for a homogeneous source are shown in Fig. 6.4 for $w < \Delta y_Q$, $w = \Delta y_Q$, and $w > \Delta y_Q$. For the problem of measuring the energy of a single line in the energy spectrum, the line

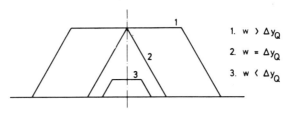

FIGURE 6.4[†]

[†] One finds $\delta T/T_0 = \max(w, \Delta y_Q)/D$.

FIGURE 6.5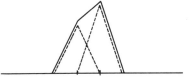

shapes recorded with $w \geqslant \Delta y_Q$ contain the same information. (The energy is determined as the mean value of the measured distribution.) With $w < \Delta y_Q$ the transmission is reduced, and therefore the time for obtaining the required quality of the data is increased.

With $w = \Delta y_Q$ the triangular line shape is obtained. The line width is given by

$$\delta T = (m\,\Delta y_P/D)T_0, \qquad (6.10)$$

and the resolving power is

$$\mathscr{R} = D/(m\,\Delta y_P). \qquad (6.11)$$

In Fig. 6.5 is shown the spectrum obtained for two sharp lines in the source spectrum when $|T_1 - T_2| = \delta T$.

The homogeneous source for which $\rho(y) = \text{const}$ so that the triangular line shape is obtained for $w = m\,\Delta y_P$ may be a rectangular area. A circular source area gives the line shape shown in Fig. 6.6; the lower

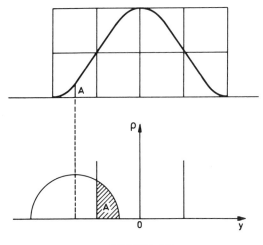

FIGURE 6.6

part of the figure shows the image distribution $\rho(y)$. For this line shape and for the triangular line shape, the peak area is the product of peak height and peak widths taken as full width at half maximum. The resolving power for a circular source is given by Eq. (6.11), when Δy_P is inserted as the source diameter.

Due to difficulties in making strong sources which are small, the source width is normally chosen such that $m \Delta y_P$ is larger than or equal to the aberration for a point source. In this sense the general expression for resolving power is given in Eq. (6.11). The aberration determines the minimum source width and thus the limit for resolving power. According to Section 1 we shall be dealing primarily with first-order focusing where aberration is not taken into account.

The transmission for the case of a point source is given by the solid angle Ω of the beam accepted for analysis. For an extended source of area (volume or length) τ, the relevant measure of transmission is $\omega = \tau \Omega$ since the usual reason for using an extended source is that the emission per unit area is limited. In some cases (e.g., Fig. 6.7) the solid angle varies in the source region, and the transmission is then given by

$$\omega = \int_\tau \Omega \, d\tau. \tag{6.12}$$

We shall now discuss the criteria for the optimal choice of parameters in a particle-analyzing system. If the emission per unit of a source area is high, a high transmission may not be required, and only a high

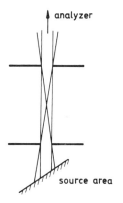

FIGURE 6.7

resolving power is necessary, whereas with a weak source, both resolving power and transmission are of importance. This may be adequately expressed by introducing a measure of "goodness" G given by

$$G = \omega \mathscr{R}, \qquad (6.13)$$

i.e., the goodness is the product of transmission and resolving power. With $\mathscr{R}$ defined by Eq. (6.11) for an extended source, a simple relationship will be derived in Section 18, namely Eq. (18.16), which relates the beam area A indicated in Fig. 18.6 to the product $\omega_y \mathscr{R}$, where ω_y is emittance in the yz projection.

When designing a particle-analyzing system, it is pertinent to establish the variation of G with the parameters of the system; the set of parameters giving the maximum value of G is the optimal choice. It should be emphasized that an optimum is usually very broad, allowing a certain measure of freedom for adapting the parameters to the special requirements of a certain experiment. Frequently it is more fruitful to investigate how $\mathscr{R}$, ω, and G vary over broad ranges of the parameters than to concentrate one's efforts to determine the exact values of parameters that give the maximum value of G.

So far, we have discussed a single-channel analyzer or a spectrometer and shall now consider the case where a row of detectors or a film is placed along the focal surface, and the field in the analyzer is kept fixed. Such a system allows the spectrum to be recorded simultaneously over a broad range of energy; the system recording the spectrum on film is called a spectrograph, while the system employing a row of detectors may be called a multichannel analyzer.

The semicircle spectrometer is shown in Fig. 6.1 and may be converted into a broad-range spectrograph by placing the beam-defining diaphragm close to the source P and a film (or another position-sensitive device) along the diameter which is the focal curve (Fig. 6.8).

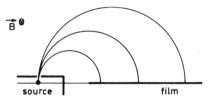

FIGURE 6.8

For a multichannel analyzer, the relevant measure of goodness G is given by

$$G = \omega N \mathscr{R},$$

where N is the number of channels. The highest number of channels to be utilized is

$$N = \Delta T/\delta T = (\Delta T/T)\mathscr{R},$$

where ΔT is the energy range; by inserting this value, we find that the goodness of a spectrograph is given by

$$G = \omega(\Delta T/T)\mathscr{R}^2, \tag{6.14}$$

When comparing a spectrometer with a spectrograph, each with the same values of $\mathscr{R}$ and G, it is noted that the transmission in the spectrometer is $\Delta T/\delta T$ times larger than the transmission in the spectrograph. Thus, the spectrometer must have good focusing properties for the central energy group, while the difficulty in the design of the spectrograph is that the resolving power must be high over the entire energy range.

3

Scaling Rules and Dispersion Coefficients

7. Scaling Rules

The concept of similarity between two systems will be introduced. This concept, and the rules of scaling are derived from nonrelativistic mechanics, but they may also be applied for relativistic velocities when the relativistic mass m_{rel} is almost constant along the particle trajectories through a system.

One can set up a reference system comprising particle and fields, in which all quantities are given the index 0. In this system, a particle (m_0, e_0) is moving. Positions are given as vectors $\mathbf{r}_0$ from a point of origin, and the time is denoted by τ_0. The motion of the particle is then described by $\mathbf{r}_0(\tau_0)$, which satisfies the equation

$$d^2\mathbf{r}_0(\tau_0)/d\tau_0^2 = (e_0/m_0)\{\mathbf{E}_0 + [d\mathbf{r}_0(\tau_0)/d\tau_0] \times \mathbf{B}_0\}, \qquad (7.1)$$

where $\mathbf{B}_0(\mathbf{r}_0)$ is the magnetic field and $\mathbf{E}_0(\mathbf{r}_0)$ the electric field, which is given by $\mathbf{E}_0 = -\operatorname{grad} V_0$ (see Section 10).

Consider now, another but similar system in which a particle (m, e) is moving. Positions are given as vectors $\mathbf{r}$, and the time is denoted by τ. The electrodes and the magnetic pole faces are obtained by the geometrical scaling $\mathbf{r} = b\mathbf{r}_0$. For all pairs of electrodes, the potential difference is given by $\Delta V = \text{const} \cdot \Delta V_0$. Thus, for each point, the field strength $\mathbf{E}$ is parallel to $\mathbf{E}_0$, and its numerical value is given by $E = \text{const} \cdot E_0$. The magnetic field has also the same shape in the two systems, and $B = \text{const} \cdot B_0$.

The concept of similarity includes the requirement that

$$\mathbf{r}(\tau) = b\mathbf{r}_0(a\tau) \tag{7.2}$$

must be a possible motion of the particle in the new system. The geometrical trajectory is scaled by the factor b, and the ratio τ/τ_0 between the times of flight for corresponding parts of the new and the old trajectory is given by $\tau/\tau_0 = a^{-1}$.

Inserting

$$d\mathbf{r}(\tau)/d\tau = ba\, d\mathbf{r}_0(\tau_0)/d\tau_0$$

and

$$d\mathbf{r}^2(\tau)/d\tau^2 = ba^2\, d^2\mathbf{r}_0(\tau_0)/d\tau_0^2$$

in the equation of motion in the new system, we obtain

$$ba^2\, d^2\mathbf{r}_0(\tau_0)/d\tau_0^2 = (e/m)\{\mathbf{E} + ba[d\mathbf{r}_0(\tau_0)/d\tau_0] \times \mathbf{B}\}. \tag{7.3}$$

It is then seen that the ratios E/E_0 and B/B_0 must be given by

$$E/E_0 = ba^2(m/m_0)e_0/e \tag{7.4}$$

and

$$B/B_0 = a(m/m_0)e_0/e. \tag{7.5}$$

In terms of potential differences, for which we have the relation

$$\Delta V/\Delta V_0 = bE/E_0,$$

Eq. (7.4) may be written

$$(\tfrac{1}{2}mv^2/e\,\Delta V) = (\tfrac{1}{2}m_0 v_0^2/e_0\,\Delta V_0), \tag{7.6}$$

where v and v_0 are the velocities.

In Eq. (7.5) we shall introduce the ratio $L/L_0 = b$ between corresponding linear dimensions, and the equation may then be written

$$mv/eBL = m_0 v_0/e_0 B_0 L_0. \tag{7.7}$$

We have now derived the scaling rules,

$$(T/e\,\Delta V) = \text{const}, \tag{7.8}$$

$$mv/eBL = \text{const}, \tag{7.9}$$

where $T = \tfrac{1}{2}mv^2$.

7. SCALING RULES

The systems may have electric fields only or magnetic fields only, but if they have both electric and magnetic fields, Eqs. (7.8) and (7.9) may be combined, and the scaling rules may then be written as

$$(m/e)\,\Delta V/B^2 L^2 = \text{const}, \tag{7.10}$$

$$vBL/\Delta V = \text{const}. \tag{7.11}$$

Consider again the reference system in which we have now introduced narrow slits defining an orbit. If the particle (m_0, e_0) is replaced by a new particle (m, e), which is found to follow the same orbit, it is seen from Eq. (7.10) that the ratios m/e and m_0/e_0 must be the same. Furthermore, Eq. (7.11) shows that the velocities v and v_0 are the same.

The ratios m/e for atomic particles may be measured in an apparatus in which the particles are emitted from a source with low velocities, then accelerated in an electric field, and finally deflected in a magnetic field. The particles enter the magnet with the energy eV_{acc}, which is well defined because of the low emission energies, $T_e \ll eV_{\text{acc}}$. The Radius R of deflection in the magnetic field B is given by

$$mv/eBR = 1, \tag{7.12}$$

and the equation determining m/e is seen to be

$$(m/e)\,V_{\text{acc}}/B^2 R^2 = 0.5. \tag{7.13}$$

Finally, we shall discuss how the scaling rules may be applied to systems with relativistic velocities. We shall consider deflection systems with a circular central path, and assume $m_{\text{rel}} \approx \text{const}$ for all trajectories. The mass, energy, and momentum are given by

$$m_{\text{rel}} = m(1 - v^2/c^2)^{-\frac{1}{2}}, \tag{7.14}$$

$$T = (m_{\text{rel}} - m)c^2, \tag{7.15}$$

and

$$p = m_{\text{rel}} v. \tag{7.16}$$

The radius of the central path is given by

$$F = m_{\text{rel}} v^2/R = 2T^*/R, \tag{7.17}$$

where F is the force, $F = eE$ or $F = evB$, and T^* is a quantity with the dimension of an energy,

$$T^* = \tfrac{1}{2} m_{\text{rel}} v^2. \tag{7.18}$$

It is seen then that we may apply scaling rules in the following form:

$$(T^*/e\,\Delta V) = \text{const}, \tag{7.19}$$

$$p/eBL = \text{const}. \tag{7.20}$$

The relativistic mass may be expressed in terms of the energy, and the mass ratio is given by

$$\mu = m_{\text{rel}}/m = 1 + T/mc^2. \tag{7.21}$$

Here, the energy T may be known from the potential. The quantity T^* is seen to be given by

$$T^* = \tfrac{1}{2}(1+\mu^{-1})T. \tag{7.22}$$

The momentum and the velocity are expressed in terms of μ by

$$p = mc(\mu^2-1)^{1/2}, \tag{7.23}$$

$$v = c(1-\mu^{-2})^{1/2}. \tag{7.24}$$

It should be mentioned that Eqs. (7.19) and (7.20) may be applied also to the lens formula for a short section of an acceleration system, when $m_{\text{rel}} \approx \text{const}$ in the section.

8. Formulas for Deflection

In Eq. (7.12), we have already given the formula for deflection in a magnetic field; however, in order to avoid tedious manipulations with units, it is useful to express this equation in a more practical form: When dealing with electrons, the charge is $e = -\varepsilon$, where $\varepsilon = 1.602 \times 10^{-19}$ C and the mass is $m = 0.911 \times 10^{-30}$ kg. In the formula for use in practice, these values are inserted, and the resulting numerical factor worked out. Similarly, when dealing with ions with mass number M and charge number n, the mass is $m = M$ amu, where 1 amu $= 1.660 \times 10^{-27}$ kg, and the charge is $e = n\varepsilon$. The values of 1 amu and ε enters the numerical factor, and the formula is expressed in terms of M and n.

The units of energy are eV (electron volt), keV, MeV, where 1 eV $= \varepsilon \cdot 1\text{V}$, and for the potential, the corresponding units V, kV, and MV are used.

For the magnetic field the units gauss and kilogauss are used. In the derivations of theoretical formula mks units are used, and for the

8. FORMULAS FOR DEFLECTION

magnetic field, the conversion from the mks unit is given by $1\text{ G} = 10^{-4}\text{ Vsec/m}^2$.

The formula obtained for magnetic deflection is

$$BR = \begin{cases} 1.066 \times 10^2 \sqrt{T}, & \text{electron} \\ 4.552 \times 10^3 (MT)^{1/2}/n, & \text{ion} \end{cases} \quad \text{cm, G, keV,} \quad (8.1)$$

and for a small deflection α obtained over the beam path length l, one has

$$\alpha = l/R = \begin{cases} 0.938 \times 10^{-2} lB/\sqrt{T} \\ 2.197 \times 10^{-4} nlB/(MT)^{1/2} \end{cases} \quad \text{cm, G, keV.} \quad (8.2)$$

It is noted that according to Eq. (7.12), the momentum is given by $p = eBR$; this is also correct for relativistic velocities. The unit gauss × centimeter is often used for the momentum, which is then related directly to the size of the magnet.

The momenta mc occurring in Eq. (7.23) take the values

$$mc = \begin{cases} 1.704 \times 10^3 \text{ G} - \text{cm,} & \text{electron} \\ M/n \cdot 3.106 \times 10^6 \text{ G} - \text{cm,} & \text{ion,} \end{cases} \quad (8.3)$$

and the energies mc^2 to be used in Eq. (7.21) are

$$mc^2 = \begin{cases} 511 \text{ keV,} & \text{electron} \\ M \cdot 931 \text{ MeV,} & \text{ion.} \end{cases} \quad (8.4)$$

For deflection in an electric field, the formula is

$$ER = 2T/n, \quad \text{cm, eV, V/cm,} \quad (8.5)$$

and small deflections are given by

$$\alpha = l/R = lnE/2T, \quad \text{cm, eV, V/cm.} \quad (8.6)$$

In Section 7, a system was described in which the ratio m/e was measured, Eq. (7.13). For ions, the ratio M/n is given by

$$M/n = 0.4826 B^2 R^2/V_{acc}, \quad \text{m, kG, MV.} \quad (8.7)$$

Rough figures for limiting factors may be given.

A magnetic field of 15–18 kG can be obtained in an electromagnet with pole pieces of soft iron. For $B = 15$ kG and $R = 100$ cm, the maximum product of mass and energy for singly charged ions is seen to be 109 amu-MeV, Eq. (8.1), so that ^{235}U ions may be deflected for $T \leqslant 0.48$ MeV.

The maximum of electric field strength between electrodes with no sharp edge is 50–100 kV/cm in normal vacuum, 10^{-6} mm Hg.

In the atmosphere outside the vacuum system (and in insulators), the maximum is lower. For instance, the length of a 0.5 MV acceleration tube must be about 120 cm as a minimum.

When an accelerator is placed inside a tank with an atmosphere of appropriate gas mixture at high pressure (~ 20 atm), the field strength may be increased so that one may obtain about 3 MV per 100 cm length of acceleration tube. Accelerators have been built with a terminal voltage going up to 7 MV.

This completes the remarks on limitations; with the above formulas for absolute quantities, the optics can be treated in the following on a relative scale.

9. Coefficients of Dispersion

In Section 2, the energy dispersion of the particle groups $(m_0, e_0, T_0 + \Delta T)$ was included in the matrix $[z_2 \leftarrow z_1]$, and according to Eq. (2.2), the outgoing ray is given by $y = [K + (z - z_2)L]\gamma$, where $\gamma = \Delta T/T_0$. Here, it is assumed that $(y_1, y_1', \gamma_1) = (0, 0, \gamma)$ and $|\gamma| \ll 1$.

With the same assumption, the coefficient of energy dispersion was introduced in Section 6, Eq. (6.2), where the dispersion was measured along the y axis at the focus for (m_0, e_0, T_0) particles,

$$y = D_T \gamma, \qquad \gamma = \Delta T/T_0. \tag{9.1}$$

If the velocity or the momentum is introduced instead of the energy, the dispersion is given by

$$y = D_v \alpha = D_p \beta, \qquad \alpha = \Delta v/v_0, \qquad \beta = \Delta p/p_0, \tag{9.2}$$

and for nonrelativistic velocities, we have the relations

$$D_v = D_p = 2D_T. \tag{9.3}$$

Already in Section 2 it was mentioned that the mass dispersion is defined as the dispersion between particles with $(m, e, T) = (m_0 + \Delta m, e_0, T_0)$, and the motivation for choosing this definition is clearly seen from Section 7. If, in the system described for the measurement of m/e, we replace the magnetic deflection by electric deflection, particles with different masses will not be separated. This follows from the fact that the mass does not occur in the scaling rule, Eq. (7.8). The mass dispersion should therefore be defined in such a way that an electric field has no mass dispersion. This gives the above condition $T = T_0$.

For deflection in a magnetic field, it is seen from Eq. (7.9) and $mv = (2mT)^{1/2}$ that the radius is a function of mT. The coefficient of mass dispersion is therefore equal to the coefficient of energy dispersion.

Thus, the coefficients for mass dispersion are given by

$$y = D_M \delta, \qquad \delta = \Delta m/m_0, \qquad |\delta| \ll 1 \qquad (9.4)$$

$$D_M = \begin{cases} 0, & \text{electric field} \\ D_T, & \text{magnetic field.} \end{cases} \qquad (9.5)$$

Finally, a charge dispersion may be introduced. Highly charged ions in slightly different charge states are present, when an ion beam of initial energy T_0 penetrates a thin foil or a gaseous target. Thus it is reasonable to define the charge dispersion as the dispersion between particles with $(m, e, T) = (m_0, e_0 + \Delta e, T_0)$. The coefficients are given by

$$y = D_n \varepsilon, \qquad \varepsilon = \Delta e/e_0, \qquad |\varepsilon| \ll 1 \qquad (9.6)$$

$$D_n = \begin{cases} -D_T, & \text{electric field} \\ -D_p = -2D_T, & \text{magnetic field.} \end{cases} \qquad (9.7)$$

These results have been derived from Eqs. (7.8) and (7.9).

Let us complete the paragraph by considering two cases, where particles moving in the z direction enter combined electric and magnetic fields.

In the first case, the electric field **E** is in the y direction and the magnetic field **B** is in the x direction, and they act over the same beam path. When $\dot{z} = -B_x/E_y$ the particles follow the z axis. If $\dot{z} \neq -B_x/E_y$, we obtain deflection in the yz plane. The system may be used for velocity selection.

In the second case both fields are in the y direction. They may act over different beam paths, and the central trajectory is curved. Here, we obtain both x and y dispersion, and such systems may be used for analysis of particles with $(m, e, T) = (m_0 + \Delta m, e_0, T_0 + \Delta T)$. For (m_0, e_0) we have $BR \propto \sqrt{T}$ and $ER \propto T$, and the trace in an xy plane will be a parabola. Instruments of this type are called parabola spectrographs.

4

Fields

10. Field Equations and Paraxial Fields

According to the framework defined in Section 1, we shall consider only static fields, and from Maxwell's equations for static electromagnetic fields, we have:

$$\text{div}\,\mathbf{E} = \rho/\varepsilon_0, \qquad \text{curl}\,\mathbf{E} = \mathbf{0},$$
$$\text{div}\,\mathbf{B} = 0, \qquad \text{curl}\,\mathbf{B} = \mu_0\mathbf{i}, \tag{10.1}$$

where $\mathbf{E}$ is the electric field strength, ρ the charge density, $\mathbf{B}$ the magnetic field strength, and $\mathbf{i}$ the current density. The value of the dielectric constant ε_0 is $(36\pi \times 10^9)^{-1}$ F/m, and the value of the permeability μ_0 is $4\pi \times 10^{-7}$ H/m. Since $\mathbf{E}$ is a rotationless field, it may be derived from a scalar potential V,

$$\mathbf{E} = -\text{grad}\,V, \tag{10.2}$$

and the differential equation for V is seen to be

$$-\text{div}\,\text{grad}\,V = \rho/\varepsilon_0. \tag{10.3}$$

The above operations may be expressed in terms of a vector operator ∇ which, in rectangular coordinates, is given by

$$\nabla = \mathbf{u}_x\,\partial/\partial x + \mathbf{u}_y\,\partial/\partial y + \mathbf{u}_z\,\partial/\partial z,$$

10. FIELD EQUATIONS AND PARAXIAL FIELDS

where $\mathbf{u}_x, \mathbf{u}_y, \mathbf{u}_z$ are unit vectors in the x, y, z directions, respectively, and the operations are given by

$$\operatorname{grad} V = \nabla V = \mathbf{u}_x \partial V/\partial x + \mathbf{u}_y \partial V/\partial y + \mathbf{u}_z \partial V/\partial z,$$

$$\operatorname{div} \mathbf{E} = \nabla \cdot \mathbf{E} = \partial E_x/\partial x + \partial E_y/\partial y + \partial E_z/\partial z,$$

$$\operatorname{div} \operatorname{grad} V = \nabla^2 V = \partial^2 V/\partial x^2 + \partial^2 V/\partial y^2 + \partial^2 V/\partial z^2,$$

$$\operatorname{curl} \mathbf{B} = \nabla \times \mathbf{B} = \mathbf{u}_x(\partial B_z/\partial y - \partial B_y/\partial z) + \mathbf{u}_y(\partial B_x/\partial z - \partial B_z/\partial x)$$
$$+ \mathbf{u}_z(\partial B_y/\partial x - \partial B_x/\partial y).$$

The equation $\operatorname{div} \mathbf{E} = \rho/\varepsilon_0$ may be integrated over the volume Ω within a closed surface Σ, and using $\int_\Omega \operatorname{div} \mathbf{E} \, d\Omega = \int_\Sigma E_n \, d\Sigma$, where E_n is the component of $\mathbf{E}$ along the outgoing normal to $d\Sigma$, we obtain Gauss's law

$$\int_\Sigma E_n \, d\Sigma = Q/\varepsilon_0, \tag{10.4}$$

where Q is the charge within Σ.

The equation $\operatorname{curl} \mathbf{B} = \mu_0 \mathbf{i}$ may be integrated over a surface Σ with a closed bounding curve s, and using $\int_\Sigma (\operatorname{curl} \mathbf{B})_n \, d\Sigma = \int_s B_s \, ds$, where B_s is the component of $\mathbf{B}$ along $d\mathbf{s}$, we obtain Ampère's law,

$$\int_s B_s \, ds = \mu_0 I, \tag{10.5}$$

where I is the current through the surface bounded by s.

We shall now consider only the outer field; i.e., the field obtained for $\rho = 0$ and $\mathbf{i} = \mathbf{0}$. For low beam intensities, the field acting on one particle may be approximated by the outer field. It is noted that for $\rho = 0$ and $\mathbf{i} = \mathbf{0}$, both $\mathbf{E}$ and $\mathbf{B}$ are rotationless and divergenceless fields, and thus the problems of determining the electric field between electrodes, and the magnetic field between pole faces are exactly similar mathematically, viz., solving the Laplace equation

$$\operatorname{div} \operatorname{grad} V = \nabla^2 V = 0 \tag{10.6}$$

for given boundary conditions. The equation is given for the electric field derived from the potential V; the magnetic field may be obtained by analogy to the electric field, and we shall not introduce a notation for the scalar field from which the magnetic field could be derived.

In the following, systems with certain symmetries will be considered, and it will be assumed that the axis potential function $V_0(z)$ defined by $V(0,0,z)$ is known. With this boundary condition to the Laplace equation, the field is fully determined. It may be expressed as a series expansion in x, y, x^2, xy, y^2, etc., with coefficients given in terms of $V_0, V_0' = dV_0/dz$, etc. For the paraxial field, only the leading terms are of importance, and these terms are easily obtained from Gauss's law with $Q = 0$.

With the symmetry referred to as *plane symmetry*, the potential is independent of y and symmetric about the yz plane,

$$V(x, y, z) = V(x, 0, z) = V(-x, 0, z). \tag{10.7}$$

It is immediately seen that $E_z = -V_0'$ and that

$$E_y = 0, \quad E_x(x, y, z) = E_x(x, 0, z) = -E_x(-x, 0, z). \tag{10.8}$$

The type of symmetry may be specified by Eq. (10.8) as well as by Eq. (10.7). The value of E_x may be determined by applying Gauss's law to the region $-x_0 \leqslant x \leqslant x_0$, $0 \leqslant y \leqslant l$, $z_0 \leqslant z \leqslant z_0 + dz$, which gives

$$2l\, dz\, E_x(x_0) + 2x_0 l[E_z(z_0 + dz) - E_z(z_0)] = 0,$$

and from this we obtain

$$E_x(x_0) = -E_z' x_0.$$

With plane symmetry, E_x and E_z are then given by

$$E_x = V_0'' x, \quad E_z = -V_0'. \tag{10.9}$$

For the axial-symmetry case, $r\varphi z$ coordinates are used, and these coordinates are introduced as follows: The y axis is replaced by the r axis for $y \geqslant 0$; rz coordinates are used in all half planes bounded by the z axis; a half-plane is specified by its angle φ from the yz plane. With the axial-field symmetry, the potential is independent of φ,

$$V(r, \varphi, z) = V(r, 0, z), \tag{10.10}$$

and it is seen that **E** lies in an rz half-plane, $E_\varphi = 0$, and that E_r is independent of φ. By applying Gauss's law to a small cylinder, we find $E_r = -\tfrac{1}{2} E_z' r$, and the paraxial field is then given by

$$E_r = \tfrac{1}{2} V_0'' r, \quad E_\varphi = 0, \quad E_z = -V_0'. \tag{10.11}$$

A magnetic field with axial symmetry,

$$B_\varphi = 0, \qquad B_r(r,\varphi,z) = B_r(r,0,z) \qquad (10.12)$$

is determined by the function $B_0(z) = B_z(0,0,z)$, and by analogy to the electric field, the paraxial field is given by

$$B_r = -\tfrac{1}{2}B_0' r, \qquad B_\varphi = 0, \qquad B_z = B_0. \qquad (10.13)$$

11. Sector, Fringing, and Quadrupole Fields

Several particle analyzers are of the sector type shown in Fig. 11.1. The central trajectory is plane, and in the field, it is a circle with radius R and center on the axis of the field sector Φ. The xyz coordinates

FIGURE 11.1

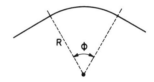

(Section 2) with z axis along the central trajectory, y axes in radial directions, and x axes normal to the yz plane, are within the sector, related to $r\varphi x$ coordinates through

$$z = z_1 + R\varphi, \qquad y = r - R, \qquad x = x. \qquad (11.1)$$

The field vectors have no components along the z axis.

A magnetic sector field of the type referred to as the $B(n)$ field is characterized in the mid-plane by $B_y = 0$ and

$$B_x = B_0 (r/R)^{-n}. \qquad (11.2)$$

For $x \neq 0$, B_y may be derived from the condition $\mathrm{curl}\,\mathbf{B} = \mathbf{0}$, which gives

$$\partial B_y/\partial x = \partial B_x/\partial y = -nB_x/r \doteq -nB_0/R$$

for $y = 0$, and from this we obtain

$$B_y = -nB_0 x/R. \qquad (11.3)$$

The electric sector field $E(n)$ is characterized in the mid-plane by $E_x = 0$ and by

$$E_y = E_0(r/R)^{-n}, \qquad (11.4)$$

and E_x for $x \neq 0$ is obtained by means of Gauss's law applied to the volume $\varphi_0 \leqslant \varphi \leqslant \varphi_0 + d\varphi$, $0 \leqslant x \leqslant x_0$, $0 \leqslant y \leqslant dr$, which gives

$$dr\, R\, d\varphi\, E_x(x_0) + x_0 R\, d\varphi\, [E_y(r) - E_y(R)] + x_0\, dr\, d\varphi\, E_y(r) = 0,$$

$$E_x(x_0) = -x_0\, dE_y/dr - x_0 E_y/R;$$

from this we obtain

$$E_x = (n-1) E_0 x/R. \qquad (11.5)$$

The field $E(1)$ is the field between coaxial cylindrical electrodes, and $E(2)$ is the field between concentric, spherical electrodes.

The homogeneous magnetic sector field $B(0)$ is, of course, obtained with plane and parallel pole faces.

For $E(1)$ and $B(0)$, the fringing fields are two-dimensional fields; such fields will now be discussed. Frequently, the fringing fields may also in other cases be approximated by two-dimensional fields. An unscreened fringing field is illustrated in Fig. 11.2. It may be the boundary region of a $B(0)$ sector, or of an $E(1)$ sector, provided R is large compared to the distance between electrodes. In the following, let us consider an electric fringing field.

In mapping a two-dimensional field, a constant potential difference δV is usually chosen between the equipotential lines. Everywhere, the distance between potential lines is given by $\delta V/E$, and since the distance

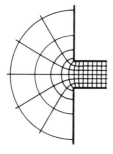

FIGURE 11.2

11. SECTOR, FRINGING, AND QUADRUPOLE FIELDS

between field lines should be proportional to E^{-1}, the two sets of lines may give a square-cell structure as may be seen in Fig. 11.2. This indicates a general method for deriving a two-dimensional field. The first sketch drawn of the field may show deviations from the square-cell structure, but by successive corrections, a fairly good impression of what the field looks like may be obtained rather quickly.

A drawing rule may also be given for the rz half-plane of a field with the axial symmetry defined by Eq. (10.10). Here, the distances between field lines and potential lines are $\delta t \propto (rE)^{-1}$ and $\delta s = \delta V/E$, respectively, and if $\delta t = \delta s$ for $r = r_0$, it follows that $\delta t/\delta s = r/r_0$ in the entire rz plane.

The particular two-dimensional field shown in Fig. 11.2 has simple structures at large distances from the boundary region. In the gap between the electrodes, the field is homogeneous (providing that the electrodes are plane, $R = \infty$), and outside the gap it takes the character of a field in a sector, π, between the two electrodes, which are half-planes with coinciding edges. It is noted that this asymptotic field may be derived from the $E(1)$ sector field between cylindrical electrodes, simply by interchanging the field lines and the potential lines. The justification of this is the symmetry between the ways the two sets of lines occur in the mapping of a two-dimensional field (i.e., the square-cell structure).

From this it is seen that at large distances, $l = z_1 - z$, the field strength in the unscreened fringing field is proportional to l^{-1}. In order to reduce the fringing field, a screening electrode is usually used as illustrated in Fig. 11.3. Its potential has the same value as that of the central trajectory, and this is usually taken to be zero.

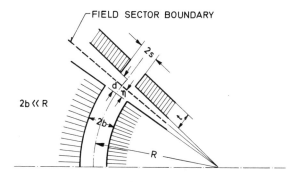

FIGURE 11.3

48　　　　　　　　　　　　　　　　　　　　4. FIELDS

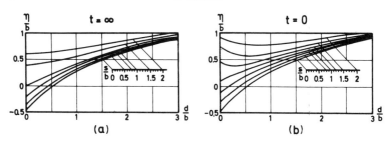

FIGURE 11.4

The effective sector boundary z_{eff} is defined by

$$\int_{-\infty}^{z_0} E_y \, dz = \int_{z_{\text{eff}}}^{z_0} E_0 \, dz, \quad (x, y) = (0, 0) \tag{11.6}$$

where z_0 is located in the interior ideal sector field, and in Fig. 11.4, calculated curves are shown from which the effective boundary may be derived for given values of the geometrical parameters indicated in Fig. 11.3.[†] It is noted that for the unscreened fringing field $\propto l^{-1}$, an effective boundary is not defined.

For obtaining a short fringing field of a sector magnet a magnetic shunt of soft iron may be used, which may be a thick plate with a long

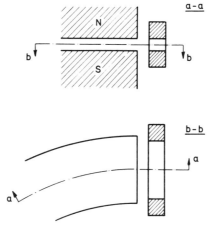

FIGURE 11.5

† H. Wollnik and H. Ewald, *Nucl. Instrum. Methods* **36**, 93 (1965).

11. SECTOR, FRINGING, AND QUADRUPOLE FIELDS   49

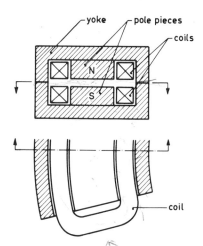

FIGURE 11.6

rectangular slit. In Fig. 11.5 two cuts *a* and *b* through pole pieces and shunt are shown. The effective field boundary may be evaluated by means of Fig. 11.4.

Sector analyzers are treated in Chapter 6 using the model with the ideal sector field within the effective boundaries and zero field outside these boundaries. The model gives correct results for first-order focusing.

Deflection magnets are normally electromagnets with the coils placed around the pole pieces so that large stray fields are avoided. Frequently, the H-type design shown in Fig. 11.6 is used. The minimum area of coil cross section is determined by the pole distance for given cooling conditions of the coils. When it is required that the fringing fields are very short and two-dimensional, the distance between the coils should be increased so that magnetic shunts may be inserted (Fig. 11.7).

FIGURE 11.7

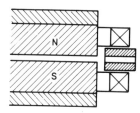

50   4. FIELDS

It will be seen later (Section 16) that certain lens actions are obtained, when the boundary of a magnetic field sector is not normal to the beam, and let us now derive a property of the fringing field, which will be used for evaluating the lens action.

As shown in Fig. 11.8, the field boundary forms an angle ε to the beam normal at the entrance point. When the center C of deflection is inside the field region (as in the figure), the angle ε is defined as positive, and for C outside the field, ε is defined as negative.

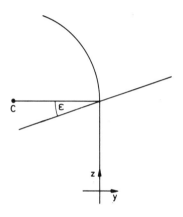

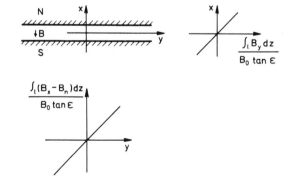

FIGURE 11.8

11. SECTOR, FRINGING, AND QUADRUPOLE FIELDS

Consider the integral $\int_l B_y \, dz$ taken along a line l parallel to the z axis, so that $x = $ const and $y = $ const. For the evaluation of this integral we shall use Ampères law $\int_s B_s \, ds = 0$, where the closed curve s is a rectangle defined as:

(1) l from z_1 in the field-free region to z_2 in the homogeneous field;
(2) from (x, y) to $(0, y)$ at z_2;
(3) from z_2 to z_1 in the midplane with $x = 0$;
(4) back to l at z_1.

When B_0 is the numerical field strength of the homogenous field, and the field has the direction opposite to the x axis, the contribution from (2) is seen to be $(2) = xB_0$. For B_z, which enters the contribution from (1), we find $B_z = -B_\perp \cos \varepsilon = -B_y \cot \varepsilon$, where $B_\perp$ is the component of field along the outgoing normal to the boundary. We then find

$$(1) = \int_l B_z \, dz = -\cot \varepsilon \int_l B_y \, dz.$$

No contributions are obtained from (3) and (4), so that $(1) + (2) = 0$, and this gives

$$\int_l B_y \, dz = xB_0 \tan \varepsilon. \tag{11.7}$$

We shall also evaluate the integral $\int_l (B_x - B_n) \, dz$, where B_n is the ideal sector field for the boundary normal to the z axis

$$B_n = \begin{cases} 0 & \text{for} \quad z < z_0 \\ -B_0 & \text{for} \quad z > z_0. \end{cases} \tag{11.8}$$

This integral is seen to be given by

$$\int_l (B_x - B_n) \, dz = yB_0 \tan \varepsilon. \tag{11.9}$$

These integrals determine the first-order optical effects of the inclined fringing field.

We now go on to describe the quadrupole field. It is a two-dimensional field not depending on z, and the xy cross section is shown in Fig. 11.9. Let us leave out derivations, and simply state the properties of the field. Around the z axis, north and south poles are arranged in alternating

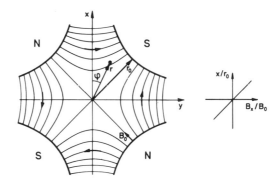

FIGURE 11.9

succession; the poles are shaped as hyperbolae given by $xy = r_0^2$, and B_0 is the numerical value of field strength at midpoints of the poles. The field lines are hyperbolae given by $x^2 - y^2 = $ const, and the components B_x and B_y of field strength are given by

$$B_x = (y/r_0) B_0, \qquad B_y = (x/r_0) B_0. \qquad (11.10)$$

It is seen that these relations are similar to the integral expressions for the inclined fringing field.

In polar coordinates $r\varphi$, one has $B_x \propto r \sin\varphi$ and $B_y \propto r \cos\varphi$, which shows that $(B_x^2 + B_y^2)^{1/2}$ is independent of φ and proportional to r.

A sextupole is shown in Fig. 11.10, and here one has $B_x \propto r^2 \sin 2\varphi$, which shows that B_x changes sign at $\varphi = 45°$, $135°$, $225°$, and $315°$, as indicated in the figure. For a $2(n+1)$ pole one has $B_x \propto \sin(n\varphi)$ for r fixed.

11. SECTOR, FRINGING, AND QUADRUPOLE FIELDS

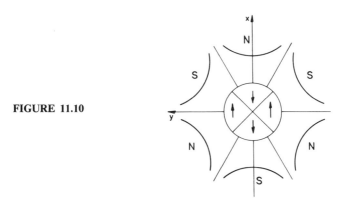

FIGURE 11.10

In Fig. 11.11 a magnetic field region with a curved boundary is shown; the boundary curve is normal to the z axis. Here, the integral $\int_l (B_x - B_n)\, dz$ is seen to be proportional to y^2, and it is noted that the same dependence on y is obtained for B_x on the y axis in a sextupole field. It may be mentioned that second-order aberration terms may be eliminated by means of a sextupole field or by means of a curved magnet boundary. However, in this book only first-order focusing is treated.

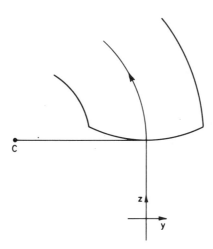

FIGURE 11.11

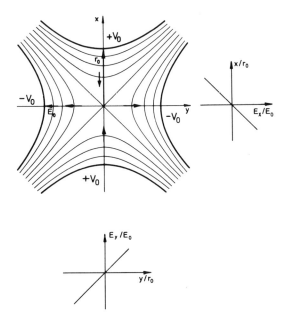

FIGURE 11.12

An electric quadrupole is shown in Fig. 11.12. The shape of the electrodes is given by $x^2 - y^2 = r_0^2$, and potential lines are indicated in the figure. The field components are given by

$$E_x = -(x/r_0) 2V_0/r_0, \qquad E_y = (y/r_0) 2V_0/r_0. \qquad (11.11)$$

It is easily verified that $V(x, y)$ obtained from Eq. (11.11) is the solution to the Laplace equation, Eq. (10.6).

5

Lenses

12. Electrostatic Lenses

In this chapter lens systems with the field symmetries dealt with in Section 10 are discussed, and first the electrostatic lens with axial symmetry is considered, Eqs. (10.10) and (10.11). It is sufficient to describe the image formation for rays in an rz plane, and here r is permitted to be negative.

The paraxial ray equation is derived from

$$\ddot{r} = r'\ddot{z} + r''\dot{z}^2$$

by inserting $\ddot{z}$, $\ddot{r}$, and $\dot{z}^2$ obtained from

$$m\ddot{z} = -eV',$$

$$m\ddot{r} = \tfrac{1}{2}eV''r,$$

$$\tfrac{1}{2}m\dot{z}^2 = -eV.$$

This gives the ray equation in which time is eliminated,

$$r'' + (V'/2V)r' + (V''/4V)r = 0; \qquad (12.1)$$

here $V = V(z)$ is the axis potential, and note that the arbitrary constant in the potential is chosen such that $T = -eV$. Frequently, V is inserted as the numerical value of the axis potential.

Lens actions are due to the radial field components occurring for

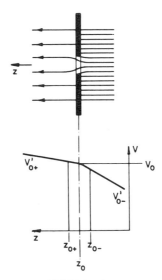

FIGURE 12.1

$V'' \neq 0$, and the simplest type lens is the aperture lens, shown in Fig. 12.1, where an electrode with a small circular aperture separates two regions with homogeneous fields in the z direction. There is only a short region $[z_{0+} \leftarrow z_{0-}]$ in which $V'' \neq 0$; this region may be treated in the thin lens approximation $r \doteq \text{const}$. From $V(z_{0+}) \doteq V(z_{0-})$, it follows that $f_s = f_i = f = -r/\Delta r'$, where $\Delta r'$ is obtained by integration of Eq. (12.1) from z_{0-} to z_{0+}. The term $V'r'/2V$ may be neglected for rays which over a long path length are paraxial. Then the integration with $V \doteq \text{const} = V_0$ gives

$$\Delta r' = -r\,\Delta V'/4V_0,$$

from which we obtain

$$1/f = (V'_{0+} - V'_{0-})/4V_0, \tag{12.2}$$

$$[z_{0+} \leftarrow z_{0-}] = \begin{bmatrix} 1 & 0 \\ (V'_{0-} - V'_{0+})/4V_0 & 1 \end{bmatrix}. \tag{12.3}$$

Note that a positive lens action $f > 0$, is obtained for $V''/V > 0$ or, in terms of the energy function $T(z)$, for $T'' > 0$, while a negative lens

12. ELECTROSTATIC LENSES

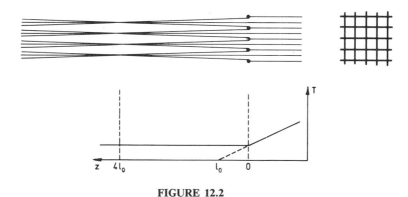

FIGURE 12.2

action $f < 0$ is obtained for $T'' < 0$. In the regions $z < z_{0-}$ and $z_{0+} < z$, where $V' = \text{const}$, the trajectories are parabolae.

Sometimes a grid is inserted in order to avoid the lens action of an aperture. The effect of having openings in the grid, which are not infinitely small, is illustrated in Fig. 12.2; the openings are referred to as facet lenses. The square openings have approximately the same lens actions as circular holes for which one has $f = 4l_0$, where $l_0 = \Delta V'/V_0 = \Delta T'/T_0$. The following deals with systems which have no grids.

Consider now a lens system with a field region between the two field-free regions with potentials V_1 and V_2. The potential function may be S-shaped for $V_1 \neq V_2$. It can be shown that with $V_1' = 0$ and $V_2' = 0$, a positive lens action is obtained.

Here, it is convenient to introduce the substitution

$$R = rV^{1/4}, \tag{12.4}$$

by means of which the ray equation may be written

$$R'' = -(3/16)(V'/V)^2 R. \tag{12.5}$$

It is noted that the substitution is chosen in such a manner that R' does not appear in the resulting equation. Since the equation shows that $R'' < 0$, the value of $R_2' - R_1'$ obtained by integration is negative; and with the proportionalities between R and r in the field-free regions, it may easily be seen that the lens action is positive.

A region with an S-shaped or bell-shaped potential function consists of regions with $T'' > 0$ and $T'' < 0$, and the variation of r may be substantial. However, since the R' term does not appear in Eq. (12.5), R

varies slowly compared with r, and for a short lens with a small variation of V, R is nearly constant. With $R \approx$ const, we obtain:

$$R_2' - R_1' \doteq -(3/16) R \int_{z_1}^{z_2} (V'/V)^2 \, dz,$$

where R_2' and R_1' are given by

$$R' = r'V^{1/4} + \tfrac{1}{4} r V' V^{-3/4}$$

at z_2 and z_1, respectively. With z_1 and z_2 placed in field-free regions, we obtain

$$1/f_2 \doteq (3/16)(V_1/V_2)^{1/4} \int_{z_1}^{z_2} (V'/V)^2 \, dz, \tag{12.6}$$

and according to Eq. (4.7), f_1 may be derived from

$$f_1/f_2 = (V_1/V_2)^{1/2}. \tag{12.7}$$

This completes the remarks in this paragraph on the electric field with axial symmetry.

The ray equation for the xz projection in the plane symmetry field, i.e., with V independent of y [Eq. (10.17)], is with $m\ddot{x} = eV''$ found to be

$$x'' + (V'/2V) x' + (V''/2V) x = 0. \tag{12.8}$$

The focal length of a slit-aperture lens is given by

$$1/f_x = (V'_{0+} - V'_{0-})/2V_0. \tag{12.9}$$

By using a substitution X, for which the term X' disappears, we obtain the weak-lens approximation

$$1/f_{2x} \doteq (7/16)(V_1/V_2)^{1/4} \int_{z_1}^{z_2} (V'/V)^2 \, dz \tag{12.10}$$

for a system between field-free regions. Such a system always has a positive lens action.

13. Acceleration System

In the following we shall consider a system consisting of the simple aperture lenses previously discussed, and a homogeneous field in which the trajectories are parabolae.

13. ACCELERATION SYSTEM

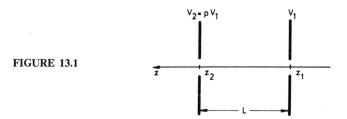

FIGURE 13.1

In Fig. 13.1 is shown an acceleration system in which we have a homogeneous acceleration field in the z direction between the two plane electrodes with small circular apertures. Outside the electrodes, the field is zero. The distance between the electrodes is $L = z_2 - z_1$, and for the potentials we have

$$V_2 = pV_1, \qquad p > 0, \tag{13.1}$$

$$V' = (p-1)V_1/L. \tag{13.2}$$

We assume that $\tfrac{1}{2}mv^2 = -eV$.

The matrix for the rz plane

$$[z_{2+} \leftarrow z_{1-}] = \begin{bmatrix} A & B \\ C & D \end{bmatrix} \tag{13.3}$$

may be obtained as

$$[z_{2+} \leftarrow z_{2-}] \begin{bmatrix} a & b \\ c & d \end{bmatrix} [z_{1+} \leftarrow z_{1-}], \tag{13.4}$$

where the matrices for the aperture lenses are obtained from Eq. (12.3).

A trajectory in the acceleration field may have $r = \text{const}$, and therefore $a = 1$, $c = 0$; it then follows from Eq. (4.6) that $d = p^{-1/2}$.

Considering next a trajectory with $r_1 = 0$, $r_1' \neq 0$, we obtain

$$b = r_2/r_1' = \dot{r}\tau/r_1' = \dot{z}_1\tau = \dot{z}_1 \int_{z_1}^{z_2} dz/\dot{z} = \int_{z_1}^{z_2} (V_1/V)^{1/2}\, dz$$

which, by means of the substitution

$$u = V/V_1, \qquad du = ((p-1)/L)\, dz$$

gives

$$b = (L/(p-1))\int_1^p du/\sqrt{u} = 2L/(p^{1/2}+1).$$

The matrix for the acceleration field is then given by

$$\begin{bmatrix} a & b \\ c & d \end{bmatrix} = \begin{bmatrix} 1 & 2L/(p^{1/2}+1) \\ 0 & p^{-1/2} \end{bmatrix}. \quad (13.5)$$

The matrices for the aperture lenses are

$$[z_{2+} \leftarrow z_{2-}] = \begin{bmatrix} 1 & 0 \\ (p-1)/4pL & 1 \end{bmatrix}, \quad (13.6)$$

$$[z_{1+} \leftarrow z_{1-}] = \begin{bmatrix} 1 & 0 \\ (1-p)/4L & 1 \end{bmatrix}. \quad (13.7)$$

By matrix multiplication, the elements of Eq. (13.3) are found to be given by

$$A = \tfrac{1}{2}(3-p^{1/2}), \qquad B = 2L/(p^{1/2}+1),$$
$$C = (3/8L)(1-p^{1/2})(1-p^{-1}), \quad D = \tfrac{1}{2}(3p^{-1/2}-p^{-1}). \quad (13.8)$$

It may be checked that $d_{12} = AD - BC = p^{-1/2}$, and it may be seen that $C < 0$ for all p's, as it should be since the lens action is positive.

In the special case of $p = 9$, we obtain

$$\begin{bmatrix} A & B \\ C & D \end{bmatrix} = \begin{bmatrix} 0 & L/2 \\ -2/3L & \tfrac{4}{9} \end{bmatrix},$$

and recalling the results from Section 3, we obtain

$$l_2/L = -(\tfrac{8}{9} - \tfrac{4}{3}l_1/L)^{-1},$$
$$m = -(\tfrac{4}{3} - 2l_1/L)^{-1},$$

where $l_1 = z_1 - z_P$ and $l_2 = z_Q - z_2$.

For $p = 1/9$, we obtain

$$\begin{bmatrix} A & B \\ C & D \end{bmatrix} = \begin{bmatrix} \tfrac{4}{3} & 3L/2 \\ -2/L & 0 \end{bmatrix},$$

$$l_2/L = \tfrac{2}{3} + \tfrac{3}{4}L/l_1,$$
$$m = \tfrac{3}{2}L/l_1.$$

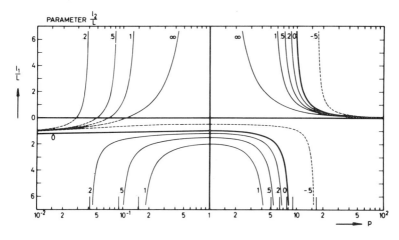

FIGURE 13.2

The general relationship between p, l_1 and l_2 is shown in Fig. 13.2. For the dotted parts of the curves, the sign of r_2 is opposite to that of r_1, so that there is a crossing in the field. The cases with no crossing may be termed "the normal cases."

With q defined as

$$q = \tfrac{8}{3}[(p^{1/2}-1)(1-p^{-1})]^{-1} \qquad (13.9)$$

one has

$$f_2 = Lq, \qquad f_1 = Lqp^{-1/2}, \qquad (13.10)$$

and the focal points and principal planes are given by

$$(H_2 - z_2)/f_2 = \tfrac{1}{2}(1-p^{1/2}), \qquad (z_1 - H_1)/f_1 = \tfrac{1}{2}(1-p^{-1/2}),$$
$$(F_2 - z_2)/f_2 = \tfrac{1}{2}(3-p^{1/2}), \qquad (z_1 - F_1)/f_1 = \tfrac{1}{2}(3-p^{-1/2}). \qquad (13.11)$$

For strong acceleration $p \gg 1$, or for strong deceleration $p \ll 1$, simple expressions may be obtained. Note, for example, that

$$z_1 - F_1 \doteq 4L/p, \qquad p \gg 1.$$

With a source point P placed at F_1, the ratio r_1/r_2 is given by

$$r_1/r_2 = \tfrac{3}{2} - \tfrac{1}{2}p^{-1/2}, \qquad P = F_1.$$

which is exact for all p.

Finally, the lens action of the system shall be described qualitatively. Let us consider a case with an acceleration lens, $p > 1$.

At the z_1 and z_2 apertures, there is a positive and a negative lens action, respectively. The radial field strength is numerically the same in the two apertures, but because of the gain in energy, the negative lens is p times weaker than the positive lens. Due to the positive lens action, a converging beam enters the acceleration field in which all trajectories are parabolae. At z_2, the beam has a smaller diameter than at z_1, and because $E_r \propto r$ [Eq. (10.11)], the forces acting on the high-energy particles at z_2 are smaller than those acting on the low-energy particles at z_1. The character of the trajectories may be described as S-shaped, and the principal planes are situated in the low-energy region. Quantitatively, it is seen from Eqs. (13.10) and (13.11) that $H_1 \doteq z_1$ and $H_2 \doteq z_1 - L/3$ for $p \gg 1$. It may be noted that $H_2 < z_1$ is a consequence of S-shaped trajectories.

The formulas, which have been derived, apply when the apertures are circular. With slit apertures, the calculations should be based on Eq. (12.9), and lens action is only obtained in the xz projection.

14. Immersion Lenses and Unipotential Lenses

The word "lens" orginates from light optics, where bodies of glass between two spherical surfaces are used. The word is therefore primarily associated with the axial-symmetry cases. Systems in which the field is two-dimensional and symmetric about a plane, are called cylinder lenses, in analogy to the corresponding components in light optics. A simple geometry with symmetry about an axis is obtained with cylindrical electrodes, but this should not be confused with cylindrical lenses.

An immersion lens has a field region situated between two field-free regions with $V_1 \neq V_2$. In a unipotential lens, the field region is situated between field-free regions with $V_1 = V_2$.

A simple immersion lens with axial symmetry consists of two

14. IMMERSION LENSES AND UNIPOTENTIAL LENSES   63

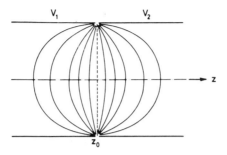

FIGURE 14.1a

cylindrical electrodes close to each other. The electrodes have identical radii r_0, and the boundary condition on the potential is

$$V(r_0) = \begin{cases} V_1, & -\infty < z < z_0, \\ V_2, & z_0 < z < \infty. \end{cases} \tag{14.1}$$

The potential distribution is shown in Fig. 14.1a. Figure 14.1b shows the variation of the axial potential Φ and its derivatives Φ', Φ'' in the

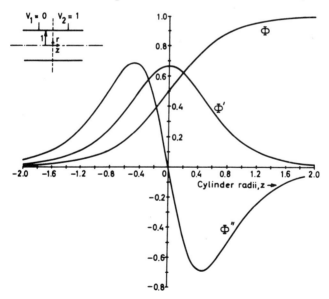

FIGURE 14.1b

special case where $V_1 = 0$ and $V_2 = 1$.[†] For the general case one has the axis potential given by $V(z) = V_1 + (V_2 - V_1) \Phi(z/r_0)$.

According to Section 12, the optical properties may be derived from the axis potential $V(z)$, and it is noted that in this case, the function has qualitatively the same shape as $V(z)$ for the system shown in Fig. 13.1. The regions where $V(z)$ is curved correspond to the aperture lenses, and in the middle we have the acceleration field. Thus, the qualitative description of the lens action has already been given in Section 13.

When $|V_2 - V_1| \ll V_1$, we have a weak lens action, and f_1 and f_2 may be obtained from Eqs. (12.6) and (12.7). Furthermore, the beam shape may be derived from Eq. (12.4) with $R \doteq$ const which, in accordance with Section 13, shows that the beam diameter is large where the energy is low.

For strong lens actions, we have no simple formulae, but in Fig. 14.2, the curves for f_2, F_2, f_1, F_1 versus V_2/V_1 are shown.

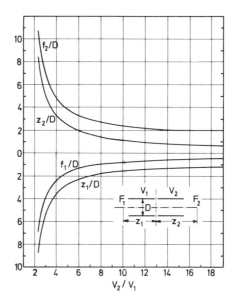

FIGURE 14.2[‡]

[†] V. K. Zworykin, G. A. Morton, E. G. Ramberg, J. Hillier, and A. W. Vance, "Electron Optics and the Electron Microscope." p. 379. Wiley, New York, 1945.

[‡] V. K. Zworykin, G. A. Morton, E. G. Ramberg, J. Hillier, and A. W. Vance, "Electron Optics and the Electron Microscope," p. 450. Wiley, New York, 1945.

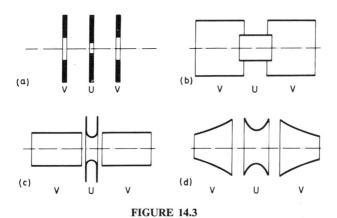

FIGURE 14.3

The treatment of aberrations is outside the scope of this book; it may be noted, however, that a lens consisting of two cylindrical electrodes is a reasonable approximation to the electrode structure giving a minimum aberration.

In Fig. 14.3, some unipotential lenses with three electrodes with potentials VUV are shown. When $|U|>|V|$, maximum energy and minimum beam diameter are obtained in the middle of the lens, while the opposite is the case when $|U|<|V|$. It may be noted in passing that suitable design of electrodes can reduce aberrations; for example, in Fig. 14.3, smaller aberrations are obtained with lenses c and d than with lenses a and b.

15. Magnetic Lenses

In Fig. 15.1, two magnetic lenses with symmetry about an axis are shown. The paraxial field is given by Eq. (10.13), and the equations of motion are

$$m\ddot{z} = -er\dot{\varphi}B_r, \tag{15.1}$$

$$m(\ddot{r}-r\dot{\varphi}^2) = er\dot{\varphi}B_z, \tag{15.2}$$

$$m(r\ddot{\varphi}+2\dot{r}\dot{\varphi}) = e(\dot{z}B_r-\dot{r}B_z). \tag{15.3}$$

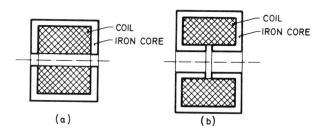

FIGURE 15.1

The last equation, combined with Eq. (10.13), shows that

$$\frac{m}{r}\frac{d}{dt}(r^2\dot{\varphi}) = -\frac{e}{2}\frac{1}{r}\frac{d}{dt}(r^2 B_z), \tag{15.4}$$

where one has $\dot{B}_z = B_z'\dot{z}$, and from this we obtain

$$\dot{\varphi} = -(e/2m)B_z - C/r^2, \tag{15.5}$$

where C is a constant of integration.

If $C = 0$, the trajectory lies in a plane containing the axis until it enters the field. Following this trajectory through the field, the outgoing ray is again lying in a plane containing the axis, but rotated through the angle φ obtained by the integration of Eq. (15.5),

$$\varphi = -(e/2m)\int B_z \, dz/\dot{z}. \tag{15.6}$$

Inserting $\dot{z}$ given by

$$\dot{z} \doteq (2T/m)^{1/2},$$

it is found that φ is given by

$$\varphi = -(e^2/8Tm)^{1/2}\int B_z \, dz. \tag{15.7}$$

The image space is rotated through the angle φ relative to the source space.

In order to find the radial ray equation, the value of $\dot{\varphi}$, given in Eq. (15.5) for $C = 0$, is inserted in Eq. (15.2). We then obtain

$$\ddot{r} = r''\dot{z}^2 + r'\ddot{z} = -(e^2/4m^2)rB_z^2. \tag{15.8}$$

Here, for paraxial rays, we may insert

$$\dot{z}^2 = 2T/m \quad \text{and} \quad \ddot{z} = 0,$$

because Eq. (15.1) with $\dot{\varphi}$ and B_r inserted shows that $\ddot{z} \propto r^2$. We thus obtain the radial ray equation

$$r'' = -(e^2/8Tm) B_z^2 r. \tag{15.9}$$

The equation shows that the magnetic field gives a positive lens action. For a short and weak lens, the focal length is given by

$$1/f \doteq (e^2/8Tm) \int B_z^2 \, dz. \tag{15.10}$$

Lens a, shown in Fig. 15.1, has $B_z = $ const in a field region with sharp boundaries, and the focal length is easily evaluated by means of Eq. (15.10). Lens b in the figure is of a type with very small aberrations, and in the electron microscope, lenses of this type are used.

16. Quadrupole and Magnetic Fringing Field Lenses

Consider a magnetic quadrupole lens. The xy cross section is shown in Fig. 11.9, and the field acts over the length L between z_1 and z_2.

Suppose that swift particles, for which only small deflections are obtained, move in the z direction. Then, x and y are nearly constant along a trajectory so that the particle moves in a constant field. The deflections $\alpha_x = x_2' - x_1'$ and $\alpha_y = y_2' - y_1'$ are given by

$$\begin{aligned} \alpha_x &= -B_y Le/(2mT)^{1/2}, \\ \alpha_y &= B_x Le/(2mT)^{1/2} \end{aligned} \tag{16.1}$$

and inserting B_y and B_x from Eq. (11.10) we find

$$\dot{\alpha}_x = -x\alpha_0/r_0, \quad \alpha_y = y\alpha_0/r_0, \tag{16.2}$$

where

$$\alpha_0 = B_0 Le/(2mT)^{1/2}. \tag{16.3}$$

Then, f_x and f_y are given by

$$f_x = r_0/\alpha_0, \quad f_y = -r_0/\alpha_0. \tag{16.4}$$

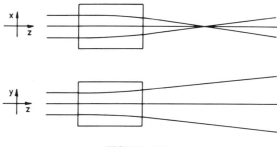

FIGURE 16.1

The principal planes are situated at $\frac{1}{2}(z_1 + z_2)$. The angle α_0 may be evaluated by means of Eq. (8.2).

The above equation illustrates that with a single quadrupole one has two distinct planes, namely a focusing plane and a defocusing plane (Fig. 16.1). It will be seen below that with combinations of quadrupole lenses with alternating field orientations, it is possible to achieve positive focusing in both planes.

Similar properties are obtained for an electric quadrupole, Fig. 11.13, for which α_0 is given by

$$\alpha_0 = E_0 Le/2T, \qquad E_0 = 2V_0/r_0. \tag{16.5}$$

The above expressions are valid for

$$\alpha_0 \ll r_0/L. \tag{16.6}$$

Suppose next that x and y vary along a trajectory through the lens. It is still assumed that $\dot{z}^2 \gg \dot{x}^2 + \dot{y}^2$ and that $\dot{z}$ is nearly constant. For x motion in a magnetic quadrupole one then has

$$\ddot{x} = -B_y e\dot{z}/m = -\omega^2 x, \qquad x = a\sin(\varphi + \eta),$$
$$\varphi = \omega t = \omega z/\dot{z} = z\Phi/L, \tag{16.7}$$

where Φ by means of Eqs. (11.10) and (16.3) is found to be given by

$$\Phi = (L\alpha_0/r_0)^{1/2}. \tag{16.8}$$

For $(x_1, x_1') = (0, x_1')$ one finds $x = a\sin\varphi$ where $a = x_1' L\Phi^{-1}$, and this gives $(x_2, x_2') = (x_1' L\Phi^{-1} \sin\Phi, x_1' \cos\Phi)$.

16. QUADRUPOLE LENSES

For $(x_1, x_1') = (x_1, 0)$ one finds $x = x_1 \cos \varphi$, which gives $(x_2, x_2') = (x_1 \cos \Phi, -x_1 L^{-1} \Phi \sin \Phi)$.

The focusing matrix is then seen to be given by

$$[z_2 \leftarrow z_1]_x = \begin{bmatrix} \cos \Phi & L\Phi^{-1} \sin \Phi \\ -\Phi L^{-1} \sin \Phi & \cos \Phi \end{bmatrix}. \tag{16.9}$$

For the y motion one obtains $y = a \sinh(\varphi + \eta)$, and the matrix is given by

$$[z_2 \leftarrow z_1]_y = \begin{bmatrix} \cosh \Phi & L\Phi^{-1} \sinh \Phi \\ \Phi L^{-1} \sinh \Phi & \cosh \Phi \end{bmatrix}. \tag{16.10}$$

When Φ is small, one finds

$$[z_2 \leftarrow z_1]_x = \begin{bmatrix} 1 - \tfrac{1}{2}\Phi^2 & L \\ -\Phi^2 L^{-1} & 1 - \tfrac{1}{2}\Phi^2 \end{bmatrix}, \tag{16.11}$$

$$[z_2 \leftarrow z_1]_y = \begin{bmatrix} 1 + \tfrac{1}{2}\Phi^2 & L \\ \Phi^2 L^{-1} & 1 + \tfrac{1}{2}\Phi^2 \end{bmatrix} \tag{16.12}$$

and this gives

$$f_x = -f_y = L\Phi^{-2} = r_0/\alpha_0 \tag{16.13}$$

in agreement with the result for swift particles.

If two quadrupole lenses are combined so that the second lens is rotated through 90° relative to the first, the arrangement is called a doublet, and one has

$$\begin{aligned} [z_3 \leftarrow z_2]_x &= [z_2 \leftarrow z_1]_y, \\ [z_3 \leftarrow z_2]_y &= [z_2 \leftarrow z_1]_x. \end{aligned} \tag{16.14}$$

For $\Phi \ll 1$ this gives

$$[z_3 \leftarrow z_1]_x = \begin{bmatrix} 1 - \Phi^2 - \tfrac{1}{4}\Phi^4 & 2L \\ -\Phi^4 L^{-1} & 1 + \Phi^2 - \tfrac{1}{4}\Phi^4 \end{bmatrix}, \tag{16.15}$$

$$[z_3 \leftarrow z_1]_y = \begin{bmatrix} 1 + \Phi^2 - \tfrac{1}{4}\Phi^4 & 2L \\ -\Phi^4 L^{-1} & 1 - \Phi^2 - \tfrac{1}{4}\Phi^4 \end{bmatrix}. \tag{16.16}$$

The focal length is now positive for both projections and is given by

$$f = L\Phi^{-4} = r_0^2/L\alpha_0^2. \tag{16.17}$$

The focal points in the image space are found to be

$$F_{ix} = z_3 - \tfrac{1}{4}L + f - (Lf)^{1/2},$$
$$F_{iy} = z_3 - \tfrac{1}{4}L + f + (Lf)^{1/2}, \tag{16.18}$$

which gives

$$F_{iy} - F_{ix} = 2(Lf)^{1/2} = 2f\Phi^2 \ll f. \tag{16.19}$$

This shows that although the focal lengths are the same in both projections, the focal points are not coincident, and this gives rise to astigmatism (Section 5). It is noted that for Φ small, the focal points are nearly coincident so that the astigmatism is small.

We have considered the doublet with the two quadrupoles placed close to each other, and obviously this gives minimum of separation between focal points.

It may be shown that for a particular source distance exact stigmatic focusing is obtained. Such a source distance may also be found, when the separation between the two quadrupoles is not zero.

The focal length for a magnetic quadrupole doublet is given by

$$f = (r_0^2/L)((2mT)^{1/2}/B_0 Le)^2 \tag{16.20}$$

and for an electric quadrupole doublet one has

$$f = (r_0^2/L)(2T/E_0 Le)^2. \tag{16.21}$$

For given values of m, e, T, r_0, and L, the same focal length is obtained when $E_0/B_0 = (2T/m)^{1/2} = \dot{z}$.

Since $f \propto r_0^2/L^3$, a strong lens is obtained for L large. The previous lenses with axial symmetry could only give strong lens action for small lens diameters. Therefore, the quadrupole doublet provides a valuable tool for focusing of high-energy particles.

In Fig. 16.2, a beam is shown in xz and yz projection. In each projection the larger beam width is obtained in the positive lens, and we may note one had a similar property for the immersion lens.

Occasionally three quadrupole lenses with alternating orientation are combined to form a triplet. The outer lenses have length L, and the middle one $2L$.

16. QUADRUPOLE LENSES

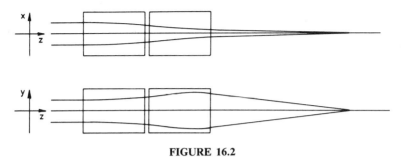

FIGURE 16.2

This arrangement may be considered as a combination of two doublets so that

$$[z_5 \leftarrow z_1]_x = [z_3 \leftarrow z_1]_y [z_3 \leftarrow z_1]_x,$$
$$[z_5 \leftarrow z_1]_y = [z_3 \leftarrow z_1]_x [z_3 \leftarrow z_1]_y, \quad (16.22)$$

where matrices on the right hand sides are given by Eqs. (16.15) and (16.16).

In Fig. 16.3, a beam is shown in xz and yz projection. Here, Φ is assumed small; a source placed at the focal point ($F_{sx} \approx F_{sy}$) of the first doublet is imaged at the focal point of the second.

Using $m = \theta_P/\theta_Q$ for magnification in each projection, it is seen that equal magnification, $m_x = m_y$, is achieved.

In the xz projection the beam shape resembles that of an unipotential lens (V, U, V) with $|U| > |V|$, and in the yz projection that obtained with $|U| < |V|$.

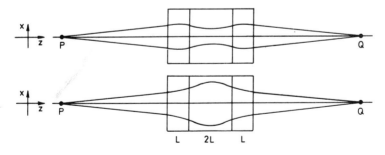

FIGURE 16.3

We now go on to consider an inclined boundary of a sector magnet, Fig. 11.8. The deflections α_x and α_y are given by Eq. (16.1), when $B_y L$ is replaced by $\int B_y \, dz = x B_0 \tan \varepsilon$ (Eq. (11.7)), and $B_x L$ is replaced by $\int (B_x - B_n) \, dz = y B_0 \tan \varepsilon$ (Eq. (11.9)). Here, B_n is the field for a boundary normal to the beam (Eq. (11.8)).

It is immediately seen that $f_x = -f_y$, and using $eB_0 R = (2mT)^{1/2}$ one finds

$$f_x = R/\tan \varepsilon, \qquad f_y = -R/\tan \varepsilon. \tag{16.23}$$

The focusing matrices are given by

$$[z_{0+} \leftarrow z_{0-}]_x = \begin{bmatrix} 1 & 0 \\ -\tan \varepsilon /R & 1 \end{bmatrix}, \tag{16.24}$$

$$[z_{0+} \leftarrow z_{0-}]_y = \begin{bmatrix} 1 & 0 \\ \tan \varepsilon /R & 1 \end{bmatrix}. \tag{16.25}$$

The angle ε is positive in Fig. 11.8, and one has $\alpha_y = y \tan \varepsilon / R$ because deflection starts at $z_0 + y \tan \varepsilon$ instead of z_0. This verifies the expression for f_y.

Positive x focusing is obtained because for $\varepsilon > 0$ the field lines in the fringing field have their concave side facing the center of deflection. This x focusing is qualitatively similar to that of a $B\,(1/2)$ sector magnet (see Fig. 19.1).

6

Analyzers

17. Electrostatic Analyzer with Cylindrical Sector Field

Sector fields of the types $E(n)$ and $B(n)$ have been treated in Section 11, and a general treatment of the sector-type analyzers will be given in Section 18. In the present section, we shall consider the special case of an analyzer with cylindrical electrodes, Fig. 17.1, which has a field of the type $E(1)$. The radial field is given by

$$E = E_0 R/r \qquad (17.1)$$

$$E_0 = -2T_0/eR. \qquad (17.2)$$

The potential is zero on the central path R and outside the sector. It is

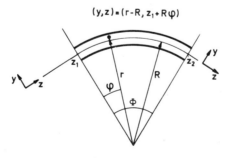

FIGURE 17.1

noted that

$$E = -2T_0/er, \qquad (17.3)$$

which shows that the energy is T_0 for all circular trajectories $r = \text{const}$ in the cylindrical field.

In the polar coordinates r, φ, the equations of motion are

$$\ddot{r} = r\dot{\varphi}^2 + (e/m)E, \qquad (17.4)$$

$$r^2\dot{\varphi} = \text{const}. \qquad (17.5)$$

For the matrix description, we shall introduce the coordinates

$$z = z_1 + R\varphi, \qquad y = r - R.$$

Furthermore, we shall introduce ω as the value of $\dot{\varphi}$ for the central path. We then obtain

$$(e/m)E = -R\omega^2 R/r, \qquad (17.6)$$

and Eq. (17.4) may be written

$$\ddot{y} = R[1+(y/R)]\dot{\varphi}^2 - R\omega^2/[1+(y/R)]. \qquad (17.7)$$

The program of this section is to derive the elements in the matrix

$$[z_2 \leftarrow z_1] = \begin{bmatrix} A & B & K \\ C & D & L \\ 0 & 0 & 1 \end{bmatrix}. \qquad (17.8)$$

We shall first determine the elements B and D by considering the trajectory with the trajectory-element at z_1 given by $(y_1, y_1', \gamma) = (0, \tan\alpha, 0)$ (cf. Section 2).

For this trajectory, the constant in Eq. (17.5) has the value $R^2\omega \cos\alpha$, and inserting the resulting value of $\dot{\varphi}$ in Eq. (17.7), we obtain

$$\ddot{y} = R\omega^2 \cos^2\alpha/[1+(y/R)]^3 - R\omega^2/[1+y/R]$$

$$= R\omega^2[(1-\alpha^2+\cdots)(1-3y/R+\cdots) - (1-y/R+\cdots)],$$

which in the first-order calculation gives the equation

$$\ddot{y} = -2\omega^2 y. \qquad (17.9)$$

By integrating Eq. (17.9), we obtain

$$y = a \sin(\sqrt{2}\,\omega t).$$

Here, the value of ωt to the first order is given by $\omega t = \varphi$. The constant a is derived from

$$y_1' = (dy/R\,d\varphi)_{\varphi=0} = a\sqrt{2}/R,$$

and thus we have found the trajectory

$$y = y_1'(R/\sqrt{2}) \sin\sqrt{2}\,\varphi. \tag{17.10}$$

The matrix elements B and D are then seen to be given by

$$B = y_2/y_1' = (R/\sqrt{2}) \sin\sqrt{2}\,\Phi, \tag{17.11}$$

$$D = y_2'/y_1' = \cos\sqrt{2}\,\Phi. \tag{17.12}$$

Our next step is to determine the elements A and C by considering a trajectory with $(y_1, y_1', \gamma) = (y_1, 0, 0)$.

As a particle enters the sector field, its energy is changed from T_0 to $T_0 - eV(y_1) = T_0 + eE_0 y_1$. Inside the sector the trajectory may intersect the central path, $y = 0$, and at this point the energy has again attained the value T_0. The trajectory may then be obtained from Eq. (17.10) by changing the phase.

It is seen that

$$y = y_1 \cos\sqrt{2}\,\varphi,$$

and

$$A = y_2/y_1 = \cos\sqrt{2}\,\Phi, \tag{17.13}$$

$$C = y_2'/y_1 = -(\sqrt{2}/R) \sin\sqrt{2}\,\Phi. \tag{17.14}$$

Finally, the elements K and L shall be determined by considering the trajectory with $(y_1, y_1', \gamma) = (0, 0, \gamma)$.

The energy of the particle in the sector field is given by

$$T(y) = T_0(1+\gamma) + eE_0 y = T_0(1+\gamma - 2y/R),$$

and $T = T_0$ is obtained for

$$y = y_\gamma = \gamma R/2.$$

6. ANALYZERS

Recalling now Eq. (17.3), it is seen that the central path for the particle group specified by γ is the circle $R_\gamma = R + y_\gamma$. Using $(y_{1\gamma}, y'_{1\gamma}, \gamma_\gamma) = (-\gamma R/2, 0, 0)$, the trajectory relative to R_γ is known. With the original central path R, the equation for the trajectory is seen to be

$$y = \gamma(R/2)(1 - \cos\sqrt{2}\varphi),$$

and we obtain the matrix elements

$$K = y_2/\gamma = (R/2)(1 - \cos\sqrt{2}\Phi), \tag{17.15}$$

$$L = y_2'/\gamma = (1/\sqrt{2})\sin\sqrt{2}\Phi. \tag{17.16}$$

The equation we have obtained for the image formation is

$$\begin{bmatrix} y_2 \\ y_2' \end{bmatrix} = \begin{bmatrix} \cos\sqrt{2}\Phi & (R/\sqrt{2})\sin\sqrt{2}\Phi \\ -(\sqrt{2}/R)\sin\sqrt{2}\Phi & \cos\sqrt{2}\Phi \end{bmatrix} \begin{bmatrix} y_1 \\ y_1' \end{bmatrix}, \tag{17.17}$$

and for the dispersion we have found

$$\begin{bmatrix} y_2 \\ y_2' \end{bmatrix} = \begin{bmatrix} (R/2)(1 - \cos\sqrt{2}\Phi) \\ (1/\sqrt{2})\sin\sqrt{2}\Phi \end{bmatrix} \gamma. \tag{17.18}$$

Let us consider for a moment the closed sector, $\Phi = 2\pi$, with a point source P emitting particles with energy T_0. Here, it follows from Eq. (17.10) that the position of the image Q is given by

$$\varphi_Q - \varphi_P = \pi/\sqrt{2} = 127.2°. \tag{17.19}$$

From this it is noted that a sector analyzer with

$$\Phi = \pi/2\sqrt{2} = 63.6°$$

has the properties

$$F_1 = z_1, \quad F_2 = z_2, \quad f = R/\sqrt{2}. \tag{17.20}$$

The relation between $l_1 = z_1 - z_P$ and $l_2 = z_Q - z_2$ is in this case found to be

$$l_1 l_2 = R^2/2. \tag{17.21}$$

When $l_1 = l_2 = R/\sqrt{2}$ is chosen, the dispersion at the focus is given by

$$y = R\gamma. \tag{17.22}$$

It is noted that for obtaining short fringing fields the distance between electrodes must be small; this, however, gives a small energy range, and the analyzer is normally used as a single-channel analyzer.

For Φ very small, a large energy range is obtained. In fact, the simple deflection system with plane electrodes may be considered as the limiting case of the sector analyzer. This system has no focusing ($f = \infty$), but nevertheless, it is in some cases well suited for particle analysis. Consider, for example, an ion beam penetrating a thin foil (Fig. 17.2).

FIGURE 17.2

All ions in the transmitted beam have about the same energy, but different charge numbers occur, $n = 1, 2, 3, ...$, and the deflection system in conjuction with a position-sensitive detector provides an efficient setup for determining relative charge-state populations.

18. Sector-Type Analyzer with Electric or Magnetic Field

The electric sector field $E(n)$, Section 11, is for the paraxial region given by

$$E_y = E_0(r/R)^{-n} \doteq E_0(1 - ny/R), \qquad E_x \doteq (n-1)E_0 x/R, \tag{18.1}$$

and the magnetic sector field $B(n)$ is given by

$$B_x = B_0(r/R)^{-n} \doteq B_0(1 - ny/R), \qquad B_y \doteq -nB_0 x/R. \tag{18.2}$$

In general, the paraxial rays in the yz plane are of the form $\sin(\varepsilon_y \varphi + \eta)$, where ε_y depends on the type of the field. For $\varepsilon_y^2 < 0$, the form $\sinh(|\varepsilon_y|\varphi + \eta)$ is obtained, but we shall be primarily concerned with cases for which $\varepsilon_y^2 > 0$. The relationship between matrix elements found for $E(1)$ in Section 17 may be generalized, and the matrix for y

focusing and energy dispersion is given by

$$\begin{bmatrix} \cos \varepsilon_y \Phi & (R/\varepsilon_y) \sin \varepsilon_y \Phi & (R/\varkappa \varepsilon_y^2)(1-\cos \varepsilon_y \Phi) \\ (-\varepsilon_y/R) \sin \varepsilon_y \Phi & \cos \varepsilon_y \Phi & (\sin \varepsilon_y \Phi)/\varkappa \varepsilon_y \\ 0 & 0 & 1 \end{bmatrix},$$ (18.3)

where, as it will be seen below, $\varkappa$ is given by

$$\varkappa = \begin{cases} 1, & E(n) \\ 2, & B(n). \end{cases}$$ (18.4)

Similarly, the rays in the xz surface are of the form $\sin(\varepsilon_x \varphi + \eta)$ for $\varepsilon_x^2 > 0$, and the matrix for x focusing is given by

$$\begin{bmatrix} \cos \varepsilon_x \Phi & (R/\varepsilon_x) \sin \varepsilon_x \Phi \\ (-\varepsilon_x/R) \sin \varepsilon_x \Phi & \cos \varepsilon_x \Phi \end{bmatrix}.$$ (18.5)

The values of ε_y and ε_x will be derived below, and the results are given in Tables 18.1 and 18.2. For $\varepsilon = 0$, no lens action is obtained, and $\varepsilon_y = \varepsilon_x$ gives stigmatic focusing.

Let us first determine ε_y for the sector magnet $B(n)$. The equation of motion is seen to be

$$\ddot{r} = r\dot{\varphi}^2 - \omega \dot{\varphi} r (R/r)^n$$

with ω given by $\omega = eB_0/m$, and for particles from the central group, we may insert $r\dot{\varphi} = R\omega$, which gives

$$\ddot{y} = -(1-n)\omega^2 y.$$

From this we obtain $\varepsilon_y = (1-n)^{1/2}$ as indicated in Table 18.2.

TABLE 18.1

	$E(n)$	$E(1)^a$	$E(2)^b$	$E(3)^c$
ε_y	$(3-n)^{1/2}$	$\sqrt{2}$	1	0
ε_x	$(n-1)^{1/2}$	0	1	$\sqrt{2}$

[a] No x focusing.
[b] Stigmatic focusing.
[c] No y focusing.

TABLE 18.2

	$B(n)$	$B(0)^a$	$B(1/2)^b$	$B(1)^c$
ε_y	$(1-n)^{1/2}$	1	$1/\sqrt{2}$	0
ε_x	$\sqrt{n}$	0	$1/\sqrt{2}$	1

[a] No x focusing.
[b] Stigmatic focusing.
[c] No y focusing.

For particles with energy $T = (1+\gamma)T_0$, it is noted that a circular orbit with radius $r_\gamma = R + y_\gamma$ is possible for $eB_x r_\gamma = (2mT)^{1/2}$, which gives

$$y_\gamma = \gamma \varepsilon_y^{-2} R/2, \qquad B(n). \tag{18.6}$$

The trajectories are then of the type $y = y_\gamma + a \sin(\varepsilon_y \varphi + \eta)$, which for $(y_1, y_1', \gamma) = (0, 0, \gamma)$ gives

$$y_2 = K\gamma = y_\gamma(1 - \cos\varepsilon_y \Phi),$$
$$y_2' = L\gamma = (y_\gamma \varepsilon_y/R) \sin\varepsilon_y \Phi, \tag{18.7}$$

and inserting y_γ, we obtain the K and L elements of Eq. (18.3).

For x focusing, the value $\varepsilon_x = \sqrt{n}$ is obtained by noting that the force in the x direction is $e\dot{z}B_y = eR\omega B_y = -n\omega^2 x/m$, which gives

$$\ddot{x} = -n\omega^2 x.$$

The special case $E(1)$ of an electric sector field has been treated in Section 17, and it is easily proved that for the general case $E(n)$ the equations of motion are

$$\ddot{y} = -(3-n)\omega^2 y, \qquad \ddot{x} = -(n-1)\omega^2 x,$$

which establish the values of ε_y and ε_x given in Table 18.1. The equilibrium orbit $r_\gamma = R + y_\gamma$ for $T = (1+\gamma)T_0$ is here given by

$$y_\gamma = \gamma \varepsilon_y^{-2} R, \qquad E(n), \tag{18.8}$$

and K and L are then obtained from Eq. (18.7). This completes the verification of Eqs. (18.3)–(18.5) and Tables 18.1 and 18.2.

The relations for image formation given in Section 3 may be utilized. With the appropriate value of ε inserted, the focusing matrix is

$$\begin{bmatrix} A & B \\ C & D \end{bmatrix} = \begin{bmatrix} \cos\varepsilon\Phi & (R/\varepsilon)\sin\varepsilon\Phi \\ (-\varepsilon/R)\sin\varepsilon\Phi & \cos\varepsilon\Phi \end{bmatrix}$$

and in terms of focus distance l_2, the magnification m and the source distance l_1 are obtained from Eqs. (3.5) and (3.8),

$$m = (\varepsilon l_2/R)\sin\varepsilon\Phi - \cos\varepsilon\Phi, \tag{18.9}$$

$$l_1 = (l_2/m)\cos\varepsilon\Phi + (R/\varepsilon m)\sin\varepsilon\Phi. \tag{18.10}$$

The quantities l_1 and l_2 are shown in Fig. 18.1.

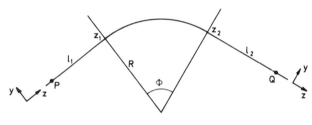

FIGURE 18.1

The coefficient D_T of energy dispersion may be calculated as $K + l_{2y}L$, but it is simpler to utilize Eq. (6.4), which with y_y inserted gives

$$D_T = \varepsilon_y^{-2}R(1+m_y)/\varkappa. \tag{18.11}$$

For detector width equal to image width $m_y \Delta y_P$, the resolving power obtained from Eq. (6.11) is seen to be

$$\mathscr{R} = T/\delta T = \varepsilon_y^{-2}R(1+m_y^{-1})/\varkappa \Delta y_P. \tag{18.12}$$

Let us consider the cases with the symmetry $l_{2y} = l_1$ and denote the common value by l. Here, the magnification m_y may be either $+1$ or -1,[†] and let us assume that $m_y = +1$. It then follows from Eq. (18.10)

[†] $m_y = -1$ may be obtained with an intermediate focus. According to Eq. (18.11) one has $D_T = 0$. For $\Phi = 2\pi/\varepsilon_y$, one obtains $m_y = -1$ with $l = 0$.

that l is given by

$$l = R(1+\cos\varepsilon_y\Phi)/\varepsilon_y \sin\varepsilon_y\Phi, \qquad (18.13)$$

and the energy dispersion $y(l)$ is found to be given by

$$y(l) = 2y_\gamma. \qquad (18.14)$$

It is noted that this dispersion does not depend on Φ. Furthermore, it is seen by means of Eqs. (18.8) and (18.6) that the dispersion $y(l) = 2R\gamma$ is obtained for $E(2)$ and $B(1/2)$, which gives stigmatic focusing, while $E(1)$ and $B(0)$ with no x focusing gives $y(l) = R\gamma$.

TABLE 18.3

	$E(1)$	$B(0)$	$E(2)$	$B(1/2)$
π/ε_y	127.2°	180°	180°	254.4°

For $\varepsilon_y\Phi = \pi$, $l = 0$ is obtained, and the resulting sector angles are given in Table 18.3. For the sector angle $\Phi = \pi/2\varepsilon_y$, the focal points are located on the sector boundaries, and it is seen that

$$l_1 l_{2y} = f_y^2 = C^{-2} = R^2\varepsilon_y^{-2}.$$

Consider now y motion with the limitation $|y| \leq d/2$. Instead of the usual yp_y plane, a modified phase plane is used, where Ry'/ε_y is plotted versus y. Having $y = a\sin(\varepsilon_y\varphi + \eta)$ for $\gamma = 0$, the phase point describes a circle, and the acceptance for $\Phi = 2\pi$ is the circular disc with radius $d/2$.

With $\Phi = \pi/2\varepsilon_y$ the acceptance is shown in Fig. 18.2 for $\gamma = 0$, and in Fig. 18.3 for the particle group with equilibrium orbit $y = y_\gamma \neq 0$. When $y_1 \geq y_\gamma$ and $y_1' \geq 0$, the maximum of y is obtained within the sector, while this is not the case, when $y_1 \geq y_\gamma$ and $y_1' < 0$. The boundary line $Ry_1'/\varepsilon_y = -d/2$ gives $y_2 = -d/2$ at the exit.

With $\Phi = \pi/2\varepsilon_y$ and $l_1 = l_{2y} = l$ one has $l = R/\varepsilon_y$. For a small source at $z_1 - l$ and a diaphragm at z_1, the beam emittance for $\gamma \neq 0$ is shown in Fig. 18.4 for $z_1 - l$, z_1, z_2, and $z_2 + l$. Also the acceptance is indicated for z_1 and z_2.

6. ANALYZERS

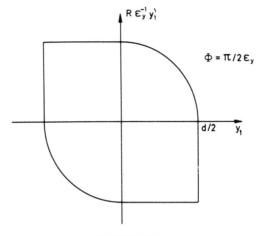

FIGURE 18.2

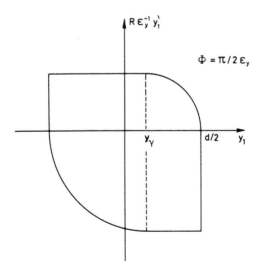

FIGURE 18.3

18. SECTOR-TYPE ANALYZER

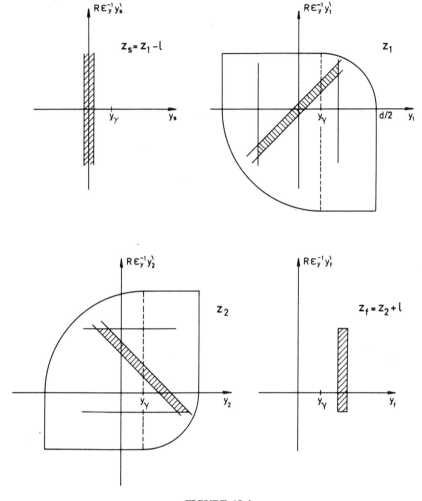

FIGURE 18.4

For a small source and $y = 0$ the emittance at z_1 is the diameter marked z_1 in Fig. 18.5, and for ϑ_1 one has $\tan \vartheta_1 = \varepsilon_y l_1/R$. The range θ_P for y' at the source is $\theta_P = \varepsilon_y (d/R) \cos \vartheta_1$. Similarly, one has $\tan \vartheta_2 = \varepsilon_y l_{2y}/R$ and $\theta_Q = \varepsilon_y (d/R) \cos \vartheta_2$. It is assumed that ϑ_1 and ϑ_2 are angles between 0 and $\pi/2$.

6. ANALYZERS

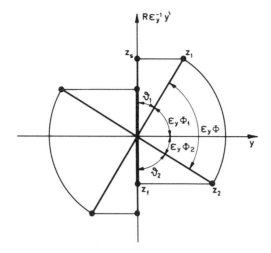

FIGURE 18.5

Using $m_y = \theta_P/\theta_Q = \cos\vartheta_1/\cos\vartheta_2$, Eq. (18.12) may be written as $\Delta y_P \theta_P T/\delta T = d\varkappa^{-1}\varepsilon_y^{-1}(\cos\vartheta_1 + \cos\vartheta_2)$. Here, $\cos\vartheta_1 + \cos\vartheta_2$ is related in a simple manner to the beam area A in the field region (Fig. 18.6). With $\varepsilon_y \Phi_1 = \pi/2 - \vartheta_1$ and $\varepsilon_y \Phi_2 = \pi/2 - \vartheta_2$ one has

$$A = Rd \int_0^{\Phi_1} \cos(\varepsilon_y \varphi)\, d\varphi + Rd \int_0^{\Phi_2} \cos(\varepsilon_y \varphi)\, d\varphi$$
$$= R\, d\varepsilon_y^{-1}(\cos\vartheta_1 + \cos\vartheta_2), \tag{18.15}$$

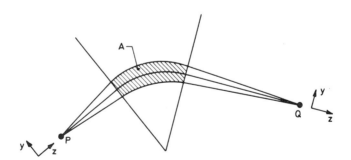

FIGURE 18.6

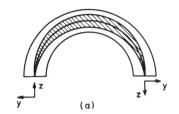

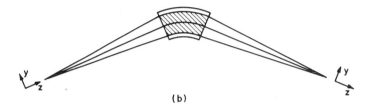

FIGURE 18.7

and this leads to the relation

$$\omega_y \mathcal{R} = A/\varkappa R, \tag{18.16}$$

where $\omega_y = \Delta y_P \theta_P$ and $\mathcal{R} = T/\delta T$.

Two magnets are shown in Fig. 18.7. They have the same area A, but the magnet in part (b) has a smaller and better utilized pole face area than the magnet in part (a).

The x motion in a magnet is limited by the pole faces, and sometimes cylindrical lenses are used as indicated in Fig. 18.8. Here, a crossover in

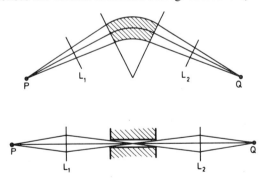

FIGURE 18.8

the middle of the magnet is obtained so that the pole distance may be small.

Quadrupole lenses (Section 16) may also be used. They affect the focusing in the yz projection, but Eq. (18.16) remains nontheless valid. This may be seen from the following: L_1 does not change the emittance ω_y; L_2 images the virtual focal plane as the resulting focal plane without changing the resolving power, since the matrices for imaging are the same for all groups within the small energy range under consideration.

When the sector boundaries of a magnet are not normal to the beam, lens actions similar to those of quadrupole lenses are obtained, and $\omega_y \mathscr{R}$ is given by Eq. (18.16). An example is shown in Fig. 18.9.

The resolving power and the transmission of an analyzer may be changed by means of preacceleration. Such a system is shown in Fig. 18.10. The energy for the central group is T_0 at the source and aT_0 in the analyzer. The analyzer images diaphragm 1 at diaphragm 3, which is placed in front of the detector. The source area is large and the beam is limited by diaphragms 1 and 2, which may have rectangular slits or circular holes. Any type of collimation may be used, i.e., slit–slit, slit–hole, hole–slit or hole–hole collimation. Diaphragm 3 has a long

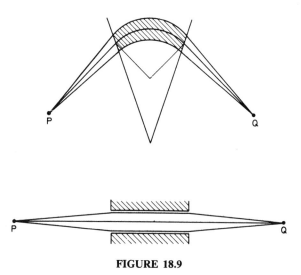

FIGURE 18.9

18. SECTOR-TYPE ANALYZER

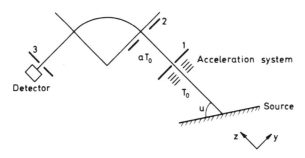

FIGURE 18.10

slit of width equal to the width of the image of 1. If 1 has a slit, the triangular line shape shown in Fig. 6.4 is obtained, and if 1 has a hole, the line shape will be as shown in Fig. 6.6.

The analyzer has the resolving power $\mathscr{R}_0 = aT_0/\delta T$, and the resulting resolving power $\mathscr{R} = T_0/\delta T$ is seen to be

$$\mathscr{R} = a^{-1}\mathscr{R}_0. \tag{18.17}$$

An increased resolving power is obtained by means of preretardation, $a < 1$.

The emittance ω_0 accepted by the collimater is

$$\omega_0 = \sigma_1 \sigma_2/l^2, \tag{18.18}$$

where σ_1 and σ_2 are the areas of diaphragms 1 and 2, which are separated by distance l. The emittance ω_{xy} of the beam entering the acceleration system is given by

$$\omega_{xy} = a\omega_0. \tag{18.19}$$

If the plane of the source forms an angle u to the z axis, the transmission ω, defined as solid angle integrated over source area, is seen to be given by

$$\omega = \omega_{xy}/\sin u. \tag{18.20}$$

Suppose now that the source area is small, and that this together with diaphragm 2 defines the beam, while diaphragm 1 is removed. For obtaining fixed focusing conditions with the source imaged at the detector slit, the acceleration system must be short, and it must be placed close to the source. Consider here a beam from a point of the source area; outside the short acceleration system this beam has the solid angle $\approx \sigma_2/l^2$, which at the source point gives the solid angle $\approx a\sigma_2/l^2$, but this shows that the above relations are approximately fulfilled.

The fixed focusing conditions are best obtained, when the lens action of the acceleration system is zero in the yz projection, i.e., the focal length is infinite. This is the case when the acceleration system has slit apertures in the y direction, which gives lens action only in the xz projection. Alternatively, fine grids may be inserted across the apertures, but here it should be recalled that the facet lenses may reduce the beam quality as illustrated in Fig. 12.2.

Suppose finally that the source is a line in the yz plane, i.e., a very narrow strip of area having the width δx; the distance from the source line to the acceleration system is large, while this system is placed close to diaphragm 1, which is imaged at the detector slit. In front of the acceleration system a Δx slit is inserted, and the diaphrams 1 and 2 have long slits; the x collimation then performed by the Δx slit and the δx source line, which determines ω_x. The slits 1 and 2 determines ω_{y0}, and ω_{xy} is given by

$$\omega_{xy} = \omega_x \sqrt{a}\, \omega_{y0}. \qquad (18.21)$$

For a source line forming the angle u to the z axis the transmission is given by Eq. (18.20) where ω_{xy} is now given by Eq. (18.21).

As mentioned previously, the resolving power can be increased by means of preretardation. With two spectral lines the spectrum recorded without preretardation may be as shown in Fig. 6.5 while the lines would be well resolved with preretardation giving $a = 0.5$. For an extended source area and beam collimation by the diaphragms 1 and 2, the product $\mathscr{R}\omega$ does not depend on a. For the above case of a system with a line source it is seen that $\mathscr{R}\omega$ is proportional to $a^{-\frac{1}{2}}$. The factor a may be an adjustable parameter, so that $\mathscr{R}$ and ω can be adapted to the requirements of various experiments.

So far energy analysis has been treated, and we may now consider mass analysis for groups $(m_0 + \Delta m, e_0, T_0)$. According to Section 9 the sector field must be a magnetic field $B(n)$. With γ replaced by $\Delta m/m_0$ and $\mathscr{R}$ defined as $m_0/\delta m$, the results obtained for energy analysis (apart from relations for preacceleration) are seen to be valid for the mass analysis.

19. Spherical Analyzer $E(2)$ and Sector Magnets $B(0)$, $B(1/2)$

The sector magnet $B(1/2)$ has conical pole faces with pole distance h given by $h = (1 + r/R)h_0/2$, and for paraxial rays, the formulae for focusing and dispersion are given in Section 18. The geometry of the pole gap is shown in Fig. 19.1. Qualitatively, the positive x focusing is obtained because the field lines have their concave side facing the center C of deflection.

The sector magnet $B(0)$ has plane pole faces and gives line focusing. An electrostatic, two-dimensional lens may be placed at the magnet entrance in order to obtain x focusing. Another possibility is to use a quadrupole lens with $f_x > 0, f_y = -f_x < 0$; here, the object point in the yz projection for the magnet is the virtual image formed by the quadrupole lens.

The relationship between l_{1y} and l_{2y} may be derived from the general formulae in Section 18, but for the magnet with homogeneous field it is simple to obtain this relationship from a geometrical consideration,

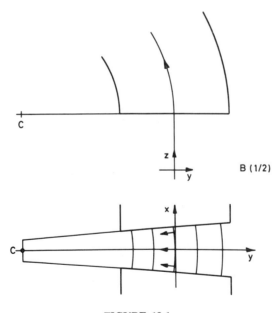

FIGURE 19.1

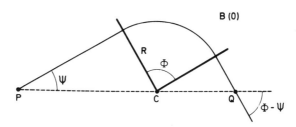

FIGURE 19.2

utilizing that all trajectories for $T = T_0$ are circles with radius R. In Fig. 19.2, P is the source point, and C is the center of the central path which is normal to the field boundaries. A trajectory from P with $y_1' \neq 0$ has the center C' of deflection, Fig. 19.3, and for y_1' small, CC' is normal to PC. Similarly, CC' is normal to PQ, where Q is the image point, and therefore Q is the intersection between the central path and the line through P and C. This simple construction is known as Barber's rule.

For $T = (1+\gamma)T_0$, the common radius radius of the orbits is $R + \Delta R$, where $\Delta R/R = \gamma/2$, and for $(y_1, y_1', \gamma) = (0, 0, \gamma)$, the outgoing ray (y_2, y_2') is given by

$$y_2 = \gamma(R/2)(1 - \cos\Phi), \qquad y_2' = (\gamma/2)\sin\Phi. \tag{19.1}$$

This may be seen either from simple geometrical considerations or from the general formulae, Eq. (18.3) and (18.4).

It is noted that the relationship between l_{1y} and l_{2y} given by Barber's rule is also obtained for the electrical sector field $E(2)$ because the value of ε_y is the same for $E(2)$ and $B(0)$, namely $\varepsilon_y = 1$.

The electric field $E(2)$ is obtained with concentric spherical electrodes,

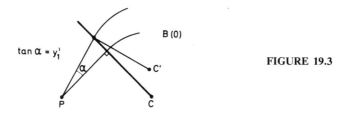

FIGURE 19.3

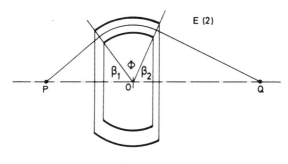

FIGURE 19.4

and the geometry of the spherical analyzer, Fig. 19.4, may be explained as follows: First two concentric circles are drawn in sector Φ of the yz plane. Then the spectrometer axis which is a line in the yz plane through the center O is chosen; it is specified by the angles β_1 and β_2 indicated in the figure ($\beta_1 + \Phi + \beta_2 = \pi$). By rotation about this axis, the circles describe the surfaces of the electrodes.

From Barber's rule it now follows that a source point P located on the spectrometer axis, Fig. 19.4, will be imaged as the point Q on the spectrometer axis, and thus the transmission in an entire "cone shell" obtained by rotation of the acceptance angle can be utilized.

For $\beta_1 = \beta_2 = \beta$, it is seen that a volume around P is imaged with magnification 1 into a volume around Q. The energy dispersion along the y axis at Q is given by $y = 2R\gamma$, as given in Section 18; the dispersion perpendicular to the spectrometer axis is then given by $r = 2R\gamma/\sin\beta$, and the dispersion along the axis by $2R\gamma/\cos\beta$.

20. Spectrographs

It is obvious that the homogeneous magnetic field inside sharp boundaries is well suited for the design of broad-range spectrographs. We have already seen one example, namely the semicircle spectrograph treated in Section 6.

In Fig. 20.1 is shown a spectrograph in which the field boundary is a circle with radius R, and for the incoming beam in a radial direction it is seen that all outgoing beams are in radial directions, i.e., normal to the boundary. Barber's rule can then be applied for determining the

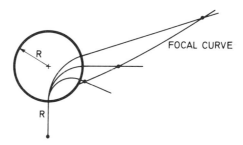

FIGURE 20.1

focal curve. For the group of particles with deflection radius R, i.e., the group with 90° deflection, it may be shown that second-order focusing is obtained. It is noted here that rays with $|y_1'|$ large have a shorter path through the magnet than they would have had if the boundaries were not curved, and it may be proved that this removes the second-order aberration of a 90° sector magnet; we leave the proof to the reader.[†] Over a range along the focal curve, including the point with second-order focusing, a high resolving power is obtained. A quadrupole lens at the magnet entrance may be used for x focusing; with $f_x = -f_y = 2R$ and source distance $2R$, the virtual source distance will still be R, while trajectories parallel to pole faces are obtained. This type of spectrograph was developed by Browne and Buechner.[‡]

A number of Browne–Buechner spectrographs may be arranged to form a magnetic circuit (like the magnets in the orange spectrometer, see Section 24), and with this instrument[§] it is possible to record spectra simultaneously for different directions of emission from a polarized source.

Another broad-range spectrograph is shown in Fig. 20.2. Here, the boundary of the homogeneous magnetic field is a straight line, and the beam direction is inclined so that it forms an angle of 54.7° to the field boundary. Recalling the definition of the ε angle given in Section 11, it is seen that $\varepsilon_1 = \varepsilon_2 = -35.3°$. It may be shown that the focal curve

---

[†] General rules for obtaining second-order focusing by means of curved field boundaries are given in E. Segré, "Experimental Nuclear Physics," Vol. I Part V. Wiley, New York, 1953.

[‡] C. P. Browne and W. W. Buechner, *Rev. Sci. Instrum.* **27**, 899 (1956).

[§] H. A. Enge. *Nucl. Instrum. Methods* **28**, 119 (1964).

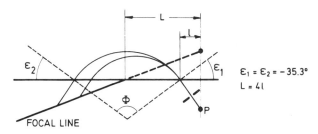

FIGURE 20.2

is a straight line, and that second-order focusing is obtained in all points. The proof of this is not complicated; it is in fact very similar to the treatment of an anlyzer with a homogeneous field given in Section 21. The inclined magnetic fringing field gives a lens action in both projections, Section 16, and it is noted that $f_x < 0$ for $\varepsilon < 0$, which reduces the transmission. It may, however, be shown that good focusing properties in the yz plane are also obtained when the sign of ε_1 is shifted (Fig. 20.3), and with ε_1 positive, the transmission is improved ($f_x > 0$).

In Section 25, the first-order focusing and dispersion of sector magnets with inclined field boundaries are derived. For a homogeneous field, a magnet of this type is specified by four parameters $(R, \Phi, \varepsilon_1, \varepsilon_2)$, while the simple magnet has only two parameters (R, Φ), and the magnet with inhomogeneous field and normal boundaries is characterized by (R, Φ, n). The magnets with inclined boundaries have many applications and are, for example, applied in parabola spectrographs in conjunction

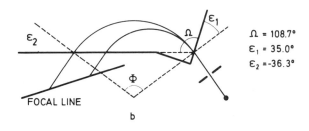

FIGURE 20.3[†]

[†] J. Borggreen, B. Elbek, and L. Perch Nielsen, *Nucl. Instrum. Methods* **24**, 1 (1963). The angles indicated in Fig. 20.3 are based on calculations to third order of aberrations.

with an electrostatic analyzer, Section 9. Here, both mass and energy analysis is performed.

In an isotope separator, the magnet is normally designed for a mass range of about ±7%, and it may be a simple sector magnet in conjunction with a lens for x focusing, a magnet with field gradient, or a magnet with inclined boundaries.

21. Analyzer with Homogeneous Electric Field

The homogeneous electric field E between plane electrodes may be used as an analyzer with focusing (shown in Fig. 21.1). The entrance to and the exit from the field are slits in the 0 electrode. Across the slits, grids must be inserted in order to avoid the lens actions of the apertures.

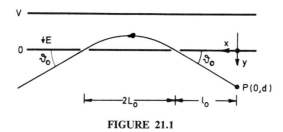

FIGURE 21.1

The central path inside the analyzer is a parabola characterized by ϑ_0 and $2L_0$ as shown in Fig. 21.1; the general parabolic orbit is characterized by corresponding parameters ϑ and $2L$.

In treating the image formation, we may use the xy coordinates where the x axis is placed along the 0 electrode and the source point P has $(x_P, y_P) = (0, d)$. We assume P outside the field and $\vartheta_0 < 45°$.†

Outside the field region the energy is $T = \frac{1}{2}Mv^2$, and in the field one has $\dot{x} = v \cos \vartheta$ and $\dot{y} = v \sin \vartheta + (t - t_1)eE/M$. The turning point of the orbit, where $\dot{y} = 0$, is reached at time t_0 given by $t_0 - t_1 = Mv \sin \vartheta / eE$. By means of $L = (t_0 - t_1) v \cos \vartheta$, the length L is then found to be given by

$$L = (T/eE) \sin 2\vartheta. \tag{21.1}$$

† The case $d = 0$, $\vartheta_0 = 45°$ is very useful and quite simple; we leave the examination of this case to the reader.

21. ANALYZER WITH HOMOGENEOUS ELECTRIC FIELD

A particle emitted from P leaves the field along the trajectory $y = (x - 2L - l)\tan\vartheta$, where $l = d\cot\vartheta$. Inserting L one finds

$$y(x, \vartheta, T) = x\tan\vartheta - (4T/eE)\sin^2\vartheta - d. \tag{21.2}$$

For $T = T_0$ the image Q is obtained where

$$\partial y/\partial\vartheta|_{\vartheta_0} = 0. \tag{21.3}$$

From this one finds

$$x_Q = (8T_0/eE)\sin\vartheta_0\cos^3\vartheta_0, \tag{21.4}$$

which by means of Eq. (21.1) may be written as

$$x_Q = 4L_0\cos^2\vartheta_0. \tag{21.5}$$

The requirement $x_Q \geqslant l_0 + 2L_0$ is fulfilled for an energy T_0 giving

$$2L_0 \geqslant l_0/\cos 2\vartheta_0. \tag{21.6}$$

The magnification is derived from the general rule $m = \theta_P/\theta_Q$, but here one simply has $\theta_Q = \theta_P = \delta\vartheta$, so that the magnification is unity,

$$m = 1. \tag{21.7}$$

For each T an image is formed, and it is noted that x and $y + d$ for the image of P are proportional to T, which shows that the focal curve is a straight line through $(0, -d)$. A second point determining the line is $(l_0 + l_0/\cos 2\vartheta_0, 0)$, which is obtained for $T = l_0 eE/\sin 4\vartheta_0$. Using the fact that $y + d$ is proportional to T one finds

$$dy/dT = (d\sin 4\vartheta_0)/l_0 eE = (\sin 4\vartheta_0)/eE \cot\vartheta_0.$$

With a source of width s normal to the beam and a detector of width $ms = s$ placed at Q the dispersion γD normal to the beam must be evaluated. Noting that $D = T_0\cos\vartheta_0(dy/dT)_{T_0}$ one finds

$$D = (T_0/eE)\sin\vartheta_0\sin 4\vartheta_0, \tag{21.8}$$

and by means of Eq. (6.11) the resolving power $\mathscr{R}$ is given by

$$\mathscr{R} = (T_0/seE)\sin\vartheta_0\sin 4\vartheta_0. \tag{21.9}$$

Let us now consider a case in which one has

$$\partial x_Q/\partial\vartheta_0 = 0. \tag{21.10}$$

With $x(y, \vartheta, T_0)$ derived from Eq. (21.2) one has $\partial x/\partial \vartheta = 0$ at Q, but in the present case one also has $\partial^2 x/\partial \vartheta^2 = 0$, so that aberration along the x axis vanished to second order. Then, the coefficient b in Eq. (5.1) is zero, which is the condition for having second-order focusing.

By means of Eq. (21.4) it is seen that Eq. (21.10) is fulfilled for

$$\vartheta_0 = 30°. \qquad (21.11)$$

It is remarkable that T_0 does not occur in the resulting condition, and this shows that second-order focusing is obtained for all points on the focal line. The points determining the line are $(0, -d)$ and $(3l_0, 0)$ where $l_0 = d\sqrt{3}$ (Fig. 21.2). An energy scale may be derived from $T = 2(d+y)eE$.

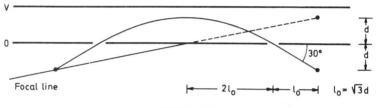

FIGURE 21.2

For $T_0 = 4\,deE$ one has $(x_Q, y_Q) = (6l_0, d)$ and the resolving power given by Eq. (21.9) is $\mathcal{R} = l_0/s$. For a point source, $\mathcal{R}$ is determined by third-order aberrations, which may be derived from $y(x, \vartheta_0 + \delta\vartheta, T_0)$ given by Eq. (21.2).

22. Coaxial Cylinder Analyzer

The analyzer is shown in Fig. 22.1. The electrodes are two coaxial cylinders with radii r_1 and r_2, and with potentials 0 and V. The radial field between the cylinders is

$$E = -[V/\ln(r_2/r_1)]r^{-1}. \qquad (22.1)$$

The inner electrode contains both entrance and exit slits; grids are inserted in order to avoid the lens action of the apertures. A point source P is placed on the z axis, i.e., the axis of the cylindrical electrodes, and the solid angle of transmission is a conical shell.

22. COAXIAL CYLINDER ANALYZER

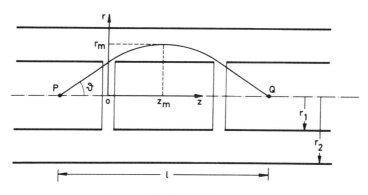

FIGURE 22.1

Each trajectory from P lies in an rz plane and is determined by the angle ϑ to the axis and the kinetic energy T. Considering a single trajectory, we may choose the zero point on the z axis where the particle enters the field. The important dimensions of the trajectory are r_m and z_m, where r_m is the maximum value of r, and z_m the corresponding value of z.

After deflection, the trajectory crosses the z axis in the point Q, for which we have

$$l = z_Q - z_P = 2(r_1 \cot \vartheta + z_m). \tag{22.2}$$

For a specified energy T, it is seen that z_m and l are functions of ϑ, and first-order focusing is then obtained for

$$\partial l / \partial \vartheta |_{\vartheta_0} = 0. \tag{22.3}$$

If also the second derivative is zero, second-order focusing is obtained.

The motion in the field is governed by the equations

$$\dot{z} = \text{const} = v \cos \vartheta, \qquad \ddot{r} = r'' v^2 \cos^2 \vartheta = eE/m.$$

Inserting the expression for the field, Eq. (22.1), we obtain

$$Ar'' = -1/2r, \tag{22.4}$$

where

$$A = A_0 \cos^2 \vartheta, \tag{22.5}$$

$$A_0 = (T/eV) \ln(r_2/r_1). \tag{22.6}$$

It may be mentioned that in Chapter 7, where space-charge effects are treated, we shall also be dealing with an equation of the type $r'' \propto r^{-1}$. A method for solving equations of this type will now be outlined.

Equation (22.4) may be written as

$$-A2r'r'' = r'/r$$

or

$$-A\, d(r'^2)/dz = d\ln r/dz,$$

and inserting the boundary conditions, we obtain

$$A(\tan^2 \vartheta - r'^2) = \ln(r/r_1). \tag{22.7}$$

The solution may be written

$$r/r_1 = \exp(\varkappa_1^2)\exp(-\varkappa^2), \tag{22.8}$$

where

$$\varkappa_1^2 = A\tan^2 \vartheta = A_0 \sin^2 \vartheta, \tag{22.9}$$

$$\varkappa^2 = Ar'^2. \tag{22.10}$$

At $z = z_m$, the value of $\varkappa$ is zero, and r_m is then given by

$$r_m = r_1 \exp(\varkappa_1^2). \tag{22.11}$$

The requirement $r_m < r_2$ is fulfilled for $\ln(r_2/r_1) > \varkappa_1^2$, and inserting the value of $\varkappa_1^2$, one has

$$eV/T > \sin^2 \vartheta. \tag{22.12}$$

The value of z_m is obtained from

$$z_m = \int_{r_1}^{r_m} dr/r',$$

where dr and r', according to Eqs. (22.8) and (22.10), are given by

$$dr = -r2\varkappa\, d\varkappa; \qquad r' = \varkappa A^{-1/2}.$$

The limits of integration r_1 and r_m correspond to $\varkappa_1$ and 0, respectively. The result is that z_m is given by

$$z_m = 2\sqrt{A}\, r_1 B, \tag{22.13}$$

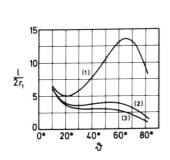
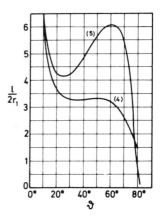

FIGURE 22.2[†]

where

$$B = F(\varkappa_1) \exp(\varkappa_1^2), \qquad (22.14)$$

$$F(\varkappa_1) = \int_0^{\varkappa_1} \exp(-\varkappa^2) \, d\varkappa. \qquad (22.15)$$

The function $F(\varkappa_1)$ is well known and is easily found in tabulated form.

It is now a matter of numerical calculation to find l as a function of ϑ. When this function has a minimum or maximum, the condition for first-order focusing is fulfilled, Eq. (22.3).

In Fig. 22.2, the curves $l/2r_1$ vs. ϑ are shown for the values of A_0 given in Table 22.1. The values of A_0 are chosen so that first-order focusing for the curves (1), (2), (4), and (5) is obtained at $\vartheta_0 = 20°$,

TABLE 22.1[a]

Curve	(1)	(2)	(3)	(4)	(5)
A_0	2.88	1.630	1.310	1.422	2.06

[a] H. Z. Sar-El, *Rev. Sci. Instrum.* **38**, 1210 (1967.)

[†] H. Z. Sar-El, *Rev. Sci. Instrum.* **38**, 1210 (1967).

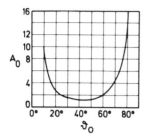

FIGURE 22.3[†]

30°, 50°, and 60°, respectively. It is clear from the figure that for each curve, there is an additional angle ϑ for which first-order focusing is obtained. For curve 3, a second-order focusing is obtained at $\vartheta_0 = 42.31°$.

In Fig. 22.3, the relation between A_0 and ϑ_0 for first-order focusing is shown. The curve has an extremum at $\vartheta_0 = 42.31°$.

The dispersion measured along the spectrometer axis is

$$\Delta z = D\gamma, \quad (22.16)$$

where

$$D = T\, \partial l(\vartheta, T)/\partial T, \quad (22.17)$$

and in Fig. 22.4 the value of D/r_1 is plotted as a function of ϑ_0.

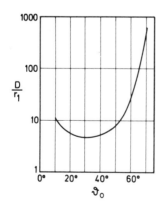

FIGURE 22.4[†]

[†]H. Z. Sar-El, *Rex. Sci. Instrum.* **38**, 1210 (1967).

23. Magnetic Lens Spectrometer

A lens with axial symmetry may be used as a spectrometer with the source and the detector situated on the z axis. The transmission in a cone shell is used so that the paraxial rays are excluded.

We shall consider the case with a homogeneous magnetic field B in the z direction. The particles (m, e) are emitted with the velocity v, and we shall determine the function $r(z)$ for $r' = \tan \alpha$ at the source. In the projection for transverse motion, the orbit is circular with the diameter $D \sin \alpha$, where

$$D = 2mv/eB \tag{23.1}$$

and the circle has its centrum situated in the distance $\tfrac{1}{2} D \sin \alpha$ from the z axis, since the source is situated on the axis. The angular frequency in this circular motion is $\dot\theta = eB/m$, which with $\dot z = v \cos \alpha$ gives

$$\theta = \dot\theta z/\dot z = 2z/(D \cos \alpha).$$

Finally, it is noted that r is given by

$$r = D \sin \alpha \sin(\theta/2),$$

which when θ is inserted gives the solution

$$r(z) = D \sin \alpha \sin(z/D \cos \alpha). \tag{23.2}$$

In Fig. 23.1, $r(z)$ is shown for $\alpha = 25°$, $30°$, and $35°$. It is seen that the beam is focused on a circle, and at this focus, the circular ring slit is placed.

High resolving power can be obtained with lens spectrometers, and they find frequent application in nuclear spectroscopy.[†]

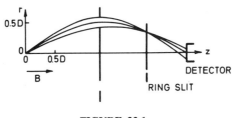

FIGURE 23.1

[†] K. Siegbahn, ed., "Alpha-, Beta-, and Gamma-Spectroscopy," Vol. I, Ch. III. North-Holland Publ., Amsterdam, 1965.

6. ANALYZERS

24. ORANGE SPECTROMETER

In the spectrometer, Fig. 24.1, six magnets form a magnetic circuit around the spectrometer axis z. The pole faces are situated in planes containing the z axis, and the field in the gaps is proportional to r^{-1}. We shall consider a family of trajectories in an rz plane, each trajectory having $r' = 0$ at $z = 0$. All the particles in the group have the same

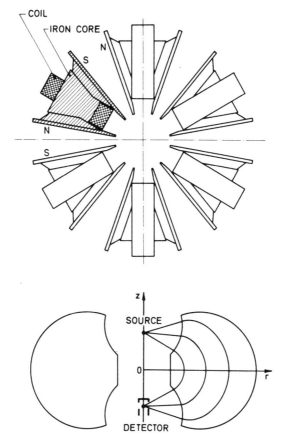

FIGURE 24.1[†]

[†] O. B. Nielsen and O. Kofoed-Hansen, *Kgl. Dan. Vidensk. Selsk. Mat.-Fys.Medd.* **29**, 6 (1955).

energy, and the trajectories are symmetrical about the plane $z = 0$. It may be shown that it is possible to determine a field boundary from which all tangents to the trajectories will cross the z axis in the same point, and at this point we may place the source. The detector is placed symmetrically to the source as shown in the figure.

It is noted that with the orange-spectrometer principle, a very high transmission may be obtained. The resolving power is also good, but due to fringing-field effects, it is lower than the resolving power in the lens spectrometers (Section 23). The orange spectrometer, however, has a higher product of resolving power and transmission. It was developed for use in high-transmission beta spectroscopy.

A further development is the iron-free orange spectrometer with electric currents flowing along the field boundary. This instrument is large, but two advantages are obtained, viz., the effect of hysteresis in the iron is avoided, so that the field is determined only by the coil current, and secondly, the fringing-field effects are eliminated to a high degree by using a large number of narrow sections.

25. Sector Magnet with Inclined Boundaries

Within the sector Ω, Fig. 25.1, the magnetic field is homogeneous. The deflection angle of the central path is Φ, and at the entrance and exit points, z_1 and z_2, respectively, the beam normal forms the angles

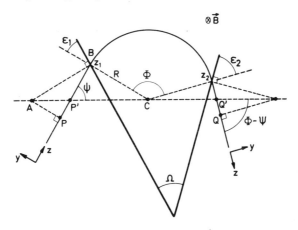

FIGURE 25.1

ε_1 and ε_2 to the field boundaries. The relation between Φ and Ω is

$$\Phi = \Omega + \varepsilon_1 + \varepsilon_2. \tag{25.1}$$

The x axis is in the field direction, and let us start by treating the x focusing.

A source point P is placed on the central path, the z axis, at the distance $l = z_1 - z_P$. Following a ray through the z_1 fringing field, we obtain from Eq. (16.24) that

$$\begin{bmatrix} x_{1+} \\ x'_{1+} \end{bmatrix} = \begin{bmatrix} 1 & 0 \\ -\tan\varepsilon_1/R & 1 \end{bmatrix} \begin{bmatrix} x_1 \\ x_1/l \end{bmatrix} = \begin{bmatrix} x_1 \\ -(x_1/R)\tan\eta \end{bmatrix},$$

where we have introduced an angle η given by

$$\tan\eta = \tan\varepsilon_1 - R/l. \tag{25.2}$$

The ray leaving the magnet is given by

$$\begin{bmatrix} x_2 \\ x_2' \end{bmatrix} = \begin{bmatrix} 1 & 0 \\ -\tan\varepsilon_2/R & 1 \end{bmatrix} \begin{bmatrix} 1 & R\Phi \\ 0 & 1 \end{bmatrix} \begin{bmatrix} x_1 \\ -(x_1/R)\tan\eta \end{bmatrix}.$$

Calculating the focus distance $l_x = -x_2/x_2'$, we find the relation

$$R/l_x = \tan\varepsilon_2 - (\Phi - \cot\eta)^{-1}. \tag{25.3}$$

For y focusing, a similar method may be applied. The matrix for $\varepsilon_1 = \varepsilon_2 = 0$, given by Eq. (18.3) for $\varepsilon_y = 1$ and $\varkappa = 2$, may be combined with the fringing-field matrices given in Eq. (16.25). The resulting matrix, including both fringing fields, is

$$[z_2 \leftarrow z_1]_y = \begin{bmatrix} \cos(\Phi-\varepsilon_1)/\cos\varepsilon_1 & R\sin\Phi \\ -\sin(\Phi-\varepsilon_1-\varepsilon_2)/(R\cos\varepsilon_1\cos\varepsilon_2) & \cos(\Phi-\varepsilon_2)/\cos\varepsilon_2 \end{bmatrix}. \tag{25.4}$$

The relation between the source distance l and the focus distance l_y could be obtained from Eq. (3.5).

We had, however, a simple construction for $\varepsilon_1 = \varepsilon_2 = 0$ (Fig. 19.2). Let us therefore determine the virtual source point P' at z_{1+}. With the angle ψ, indicated in Fig. 25.1, the distance to P' is given by

$$l_{P'} = R\cot\psi,$$

25. SECTOR MAGNET WITH INCLINED BOUNDARIES

and by applying the fringing-field matrix, Eq. (16.25) we obtain

$$\begin{bmatrix} y_1 \\ y_1/l_{P'} \end{bmatrix} = \begin{bmatrix} y_1 \\ (y_1/R)\tan\psi \end{bmatrix} = \begin{bmatrix} 1 & 0 \\ \tan\varepsilon_1/R & 1 \end{bmatrix} \begin{bmatrix} y_1 \\ y_1/l \end{bmatrix}.$$

From this it is seen that ψ is given by

$$\tan\psi = \tan\varepsilon_1 + R/l, \tag{25.5}$$

and Eq. (25.5) shows that P' is obtained by the construction indicated in Fig. 25.1. The intersection point between the beam normal from P and the normal to the field boundary at the entrance point B is A. The line AC cuts BP in P'. A similar construction is applied for the exit side for obtaining the real image Q from the virtual image Q'.

The construction is named Cartan's construction.

From a relation analogous to Eq. (25.5) it is seen that the focus distance l_y is given by

$$R/l_y = \tan(\Phi - \psi) - \tan\varepsilon_2. \tag{25.6}$$

Using the construction, the focal curve over a broad energy range may be determined fairly quickly.

In the first-order calculation, the central ray in the dispersed beam is given by $(y_2, y_2') = (K\gamma, L\gamma)$. This may be obtained from Eq. (19.1) by adding to y_2' the "lost" deflection $y_2 \tan\varepsilon_2/R$. The resulting expressions are

$$K = (R/2)(1 - \cos\Phi) = (R/2)\sin\Phi\tan(\Phi/2),$$
$$L = \tfrac{1}{2}\sin\Phi[1 + \tan\varepsilon_2\tan(\Phi/2)]. \tag{25.7}$$

The matrix for image formation for the group $T_0(1+\gamma)$ is obtained from Eq. (25.4) by means of the corrections $R \to R + \tfrac{1}{2}\gamma R$ and

$$\Phi \to \Phi - y_2' = \Phi - L\gamma, \qquad \varepsilon_2 \to \varepsilon_2 - y_2' = \varepsilon_2 - L\gamma.$$

These first-order corrections may also be applied in Eqs. (25.3) and (25.6).

The important property obtained by using inclined-field boundaries is the x focusing which may improve the transmission. Point focusing may be obtained, and Eqs. (25.3) and (25.6) show that the condition for $l_x = l_y$ is

$$\tan(\Phi - \psi) - 2\tan\varepsilon_2 = (\cot\eta - \Phi)^{-1}. \tag{25.8}$$

Normally, astigmatism occurs for $\gamma \neq 0$. (Astigmatism is avoided over an energy range when the tangents to the focal curves are coinciding. For $l = \infty$, this is obtained when $(\cot \varepsilon_1 - \Phi)(\cot \varepsilon_2 - 2 \tan \varepsilon_2) = 4$, and Eq. (25.8) are fulfilled.)

We shall now consider the symmetric cases with

$$\varepsilon_1 = \varepsilon_2 = \varepsilon, \qquad l_x = l_y = 1. \tag{25.9}$$

Here, the angles $\Phi - \psi$ and ψ have the same value,

$$\Phi - \psi = \psi = \Phi/2, \tag{25.10}$$

and Eq. (25.5) gives

$$\tan(\Phi/2) = \tan \varepsilon + R/l. \tag{25.11}$$

The case $l = l_x$ may be realized with trajectories for which $x = \text{const}$ in the magnet, and in this case we have $l = f_x$, from which $\tan \varepsilon = R/l$, or $\eta = 0$, is obtained. This, together with Eqs. (25.8) and (25.10), shows that the conditions for the symmetry are

$$\tan \varepsilon = R/l = \tfrac{1}{2} \tan(\Phi/2). \tag{25.12}$$

A calculation of the dispersion at the focus gives

$$y = 2R\gamma. \tag{25.13}$$

The equation $l = l_x$ may also be fulfilled with a cross-over, $x = 0$, in the middle of the magnet. In this case we have

$$\tfrac{1}{2}R\Phi = -x_{1+}/x_{1+}' = R/\tan \eta$$

or

$$\Phi \tan \eta = 2.$$

When this is combined with Eqs. (25.2) and (25.11), we find the conditions for the symmetry with crossing,

$$\begin{aligned}\tan \varepsilon &= \tfrac{1}{2}[\tan(\Phi/2) + 2/\Phi], \\ R/l &= \tfrac{1}{2}[\tan(\Phi/2) - 2/\Phi].\end{aligned} \tag{25.14}$$

Solutions with $l \geqslant 0$ are found for $\Phi \geqslant 98.6°$. It may be noted that

$$\begin{aligned} l &= l_x = l_y = \infty, \\ \varepsilon &= \Phi/2, \qquad \Omega = 0, \end{aligned} \qquad \text{for } \Phi = 98.6°.$$

25. SECTOR MAGNET WITH INCLINED BOUNDARIES

When $\Omega = 0$, it is a general feature that $l_y = \infty$ is obtained for $l = \infty$, which may be seen from Cartan's construction. For $\Phi > 98.6°$, the dispersion at focus is found to be $y = 3R\gamma$. for the symmetry with cross-over.

When Eq. (25.9) holds, it is trivial to show that the magnifications m_x, m_y, m_z are unity. If the incoming beam from a point source has a circular cross section, the outgoing beam will also have a circular cross section.

Of interest is also the case where the incoming beam is a parallel beam $l = \infty$, with a circular cross section. The ratio $e = y_2/x_2$ is determined by Φ and ε_1. Expressing ε_1 in terms of e and Φ, we obtain the relation

$$\tan \varepsilon_1 = (e - \cos \Phi)/(e\Phi + \sin \Phi). \quad (25.15)$$

If $(\Phi, \varepsilon_1, \varepsilon_2)$ fulfils the condition in Eq. (25.8) for point focusing, the ratio between the magnifications m_y and m_x is given by e. The relationship between e, ε_1, ε_2, and Φ is shown in Fig. 25.2. On the dotted curve, the tangents to the focal curves for x and y focusing coincide.

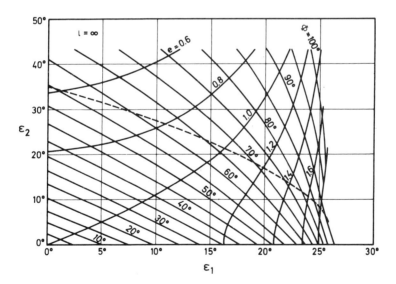

FIGURE 25.2

Besides the applications mentioned in Section 20, inclined magnetic field boundaries are used for sector focusing in cyclic particle accelerators.† A few remarks on this topic may be included.

It is recalled that the inclined-magnet boundary has an effect similar to the quadrupole lens, and that a quadrupole pair gives positive lens action in both projections. From this it is seen that the arrangement shown in Fig. 25.3 may be used for focusing. The figure illustrates part of a cyclic acclerator.

FIGURE 25.3

The principle in a different version is used in the sector-focused cyclotron. Before describing this, we will briefly outline the principle of the classical cyclotron and its dilemma.

Between the plane and parallel pole faces, two electrodes are placed. Their geometrical shape may be obtained by dividing a flat cylindrical box along a diameter.

Let a particle describe a semicircle, $eBR = mv$, in electrode 1 and then be accelerated in the gap when passing from 1 to 2. While it is describing the semicircle $R + \Delta R$ in 2, the sign of the potential difference between the electrodes is changed; the particle will be accelerated again when passing from 2 to 1. This is repeated in many revolutions.

The particle describes a semicircle in the time

$$\tfrac{1}{2}\tau = \pi R/v = \pi m/eB.$$

For nonrelativistic velocities and homogeneous field, the potential difference should be changed with the constant frequency

$$1/\tau = eB/2\pi m.$$

This is called the cyclotron frequency.

However, when the particle becomes relativistic, the mass increases. Therefore B should increase with R, but that would cause a defocusing,

† Livingood, "Principles of Cyclic Particle Accelerators." Van Nostrand-Reinhold, Princeton, New Jersey, 1961

25. SECTOR MAGNET WITH INCLINED BOUNDARIES

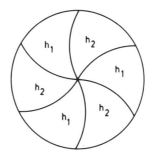

FIGURE 25.4

and the particle could not go on circling around in the flat box. (This applies even though a focusing is obtained each time the particle crosses the acceleration gap.)

We may now illustrate the principle of sector focusing. In Fig. 25.4, the pole face is divided into sectors by curves which, for increasing R, form increasing angles to the radial directions. From sector to sector, the pole distance changes between h_1 and h_2. Thus, the field is changing because $B_1 h_1 = B_2 h_2$. The lens action may be derived by the method outlined in Section 11, where the integral now has a contribution from (4). By the alternating lens actions, focusing is obtained as illustrated in Fig. 25.5. The focusing principle is frequently termed "strong focusing".

With the use of sector focusing, the pole distances h_1 and h_2 may be reduced for increasing R, and it may then be understood that a constant frequency may be obtained.

FIGURE 25.5

7

Space Charge and Beam Production

26. Ideal Beam in a Drift Region

For a high-intensity beam, particle optics is essentially different from the optics previously outlined since interactions between particles in the beam are of significance. In this chapter, the effect of mutual electrostatic repulsions is treated, while magnetic interactions are not taken into account. The magnetic interactions are only important at very high particle velocities.

In the space-charge model, the force on one particle, due to electrostatic repulsions from other particles, is derived from a continuous distribution of charge in the beam, and this distribution must be consistent with the set of resulting trajectories. The model gives the proper beam structure averaged over time. Here, only steady beams are considered.

The space-charge effect will be treated for a narrow beam through a drift region, i.e., a region with no outer field. The kinetic energy of a particle is mainly due to the velocity component in the beam direction, and a monoenergetic beam may be considered. Using numerical values for charge e and for the potential V of the drift region, the particle energy is given by $T \doteq \tfrac{1}{2}m\dot{z}^2 \doteq eV$.

The space-charge effect will be treated for the following two ideal-beam structures:

(i) *Ideal Round Beam.* This beam has a circular cross section with radius $R = R(z)$. At each z, the current density is homogeneous over the cross section, and the set of trajectories, in the $r\varphi z$ coordinates, is characterized by $\varphi' = 0$ and $r' \propto r$. It will be seen below that these conditions are fulfilled for all z when they are fulfilled for one z. Each trajectory lies in a plane containing the z axis, and the tangents at z determine a virtual focus point on the z axis at distance $l = -r/r' = -R/R'$ from z; $l \gg R$. The angle of the beam convergence at z is 2β, where $\beta = -R' = R/l$. It will be seen below that the laminar structure is conserved; a crossover does not exist, and at a certain z value, z_m, the beam radius has a minimum value R_m. For R and β specified at z, we shall determine R_m and the distance $l_m = z_m - z$. In Fig. 26.1 are indicated R, β, l, R_m, l_m, and an angle β_m defined as $\beta_m = R/l_m$.

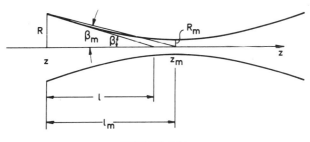

FIGURE 26.1

(ii) *Ideal Flat Beam.* This is a two-dimensional beam of infinite height, $-\infty \leqslant x \leqslant \infty$, and of width $2Y$, $|y| \leqslant Y(z)$. The current density is homogeneous over each beam cross section, and the trajectories have $x' = 0$ and $y' \propto y$. The convergence is 2β, whereas $\beta = -Y'$, and the distance to the virtual focus is $l = -Y/Y' = Y/\beta$. It will be seen below that a true crossover exists at low intensities, while the crossover is avoided above a certain intensity limit.

Consider the ideal round beam. Since it is a nearly cylindrical beam ($l \gg R$), the axial component of the charge field is small, $E_z \doteq 0$, and the radial component E_r is derived by applying Gauss's law to the core with radius $r \leqslant R$. When N is the number of particles in the beam section between z and $z + \Delta z$, the charge in the core is $eN\pi r^2/\pi R^2$, so that

$$2\pi r \, \Delta z \, E_r = \varepsilon^{-1} e N r^2 / R^2.$$

Introducing the beam current I given by

$$I = eN\dot{z}/\Delta z = (eN/\Delta z)(2eV/m)^{1/2},$$

we find

$$E_r = (rI/R^2 2\pi\varepsilon)(m/2eV)^{1/2}.$$

The equation of motion is

$$m\ddot{r} = 2eVr'' = eE_r,$$

which with E_r inserted may be written as

$$r'' = Cr/R^2,$$

where the dimensionless constant is given by

$$C = (4\pi\sqrt{2}\varepsilon)^{-1} m^{1/2} I/e^{1/2} V^{3/2}. \tag{26.1}$$

For trajectories at the surface of the beam section at z, we find $R'' = C/R$, and inside the surface

$$r'' = C_c/r, \quad \text{where} \quad C_c = C\pi r^2/\pi R^2$$

It is seen that $r'' = R''r/R$, which shows that the condition $r' \propto r$ is maintained for all z, and with this laminar structure, C_c is a constant for a core of the beam. Thus $r'' = C_c/r$ is the ray equation, and the differential equation for the beam radius is

$$R'' = C/R. \tag{26.2}$$

For an electron beam, $P = I\dot{V}^{-3/2}$ is known as the *perveance*, in terms of which C is given by

$$C = 4.80 \times 10^{-4} P, \quad P = (I \text{ mA})(V \text{ kV})^{-3/2}, \quad \text{electrons.} \tag{26.3}$$

For an ion beam, the mass number M and the charge number n enter the relevant measure of perveance, and C is given by

$$C = 2.05 \times 10^{-5} P, \quad P = (M/n)^{1/2} (I \, \mu\text{A})(V \text{ kV})^{-3/2}, \quad \text{ions} \tag{26.4}$$

These relations are used for scaling in space-charge problems, and C may be called the normalized perveance.

It is obvious from the ray equation that a crossover does not exist, and R_m and z_m will now be derived for a given value of C, and for $(R, \beta) = (R_1, \beta_1)$ at $z = z_1$.

Equation (26.2) for R is solved by using the method outlined in Section 22, and from this reference we may immediately give the results. A parameter $\varkappa_1$ is defined as

$$\varkappa_1 = -R_1'/(2C)^{1/2} = \beta_1/(2C)^{1/2}, \tag{26.5}$$

and R_m and $l_{m1} = z_m - z_1$ are given by

$$R_m = R_1 \exp(-\varkappa_1^2), \tag{26.6}$$

$$l_{m1} = l_1 2\varkappa_1 \exp(-\varkappa_1^2) F(\varkappa_1), \tag{26.7}$$

where $l_1 = R_1/\beta_1$ and

$$F(\varkappa_1) = \int_0^{\varkappa_1} \exp(\varkappa^2) \, d\varkappa. \tag{26.8}$$

In Section 22 we obtained the more familiar function with the integrand $\exp(-\varkappa^2)$, but also the function defined in Eq. (26.8) is tabulated.[†]

The ratios R_m/R_1 and l_{m1}/l_1 are seen to be functions of $C\beta_1^{-2}$, and these functions are shown in Fig. 26.2. Furthermore, l_{m1}/l_1 is plotted versus R_m/R_1 in Fig. 26.3, which also shows the curve for $C\beta_{m1}^{-2}$ versus R_m/R_1, where $\beta_{m1} = R_1/l_{m1}$ so that $C\beta_{m1}^{-2} = (l_{m1}/l_1)^2 C\beta_1^{-2}$.

For fixed values of β_1 and R_1, and thus for l_1, it may be seen from Fig. 26.2 how the beam shape depends on C. For C increasing from 0 to ∞, R_m increases from 0 to R_1, while l_{m1} first increases from l_1 to a maximum and then decreases to 0. The cases with maximum of l_{m1}, and with $l_{m1} = l_1$ on the decreasing branch, are characterized by

$$\begin{aligned} l_{m1}/l_1 &= 1.285, & 1, \\ R_m/R_1 &= 0.105, & 0.426, \\ C\beta_1^{-2} &= 0.222, & 0.585, \\ C\beta_{m1}^{-2} &= 0.367, & 0.585. \end{aligned} \tag{26.9}$$

An alternative means of specifying beam shape is to define R_m and C. The shape may be described by the function $l_m(R)$, and two methods for

[†] K. A. Karpov, "Tables of the Function $F(z) = \int_0^z e^{x^2} dx$," Mathematical Tables Series, Vol. 23. Pergamon, Oxford, 1964; K. A. Karpov, "Tables of the Function $W(z) = e^{-z^2} \int_0^z e^{x^2} dx$ in the Complex Domain," Mathematical Tables Series, Vol. 27. Pergamon, Oxford, 1965.

7. SPACE CHARGE AND BEAM PRODUCITON

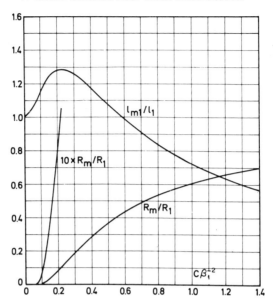

FIGURE 26.2

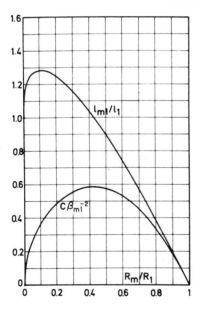

FIGURE 26.3

deriving $l_m(R)$ may be outlined as follows: (i) From Eq. (26.6), here written as

$$\beta = [2C \ln(R/R_m)]^{1/2}, \quad (26.10)$$

one obtains $\beta(R)$. With this inserted in $l = R/\beta$ and in $C\beta^{-2}$, which by means of Fig. 26.2 determines l_m/l, the beam shape $l_m(R)$ is obtained as the product of l and l_m/l. (ii) From the curve in Fig. 26.3 for $C\beta_m^{-2}$ versus R_m/R, the function $\beta_m(R)$ is obtained, and the beam shape $l_m(R)$ is then obtained from $l_m = R/\beta_m$.

Frequently, it is useful to specify a beam by R_m and the radius R_1 at a given distance l_{m1} from z_m. Here, $\beta_{m1} = R_1/l_{m1}$ is known, and C may then be determined by means of the curve for $C\beta_{m1}^{-2}$ versus R_m/R_1. It is noted that $C\beta_{m1}^{-2}$ has a maximum for $R_m/R_1 = 0.426$, where $l_1 = l_{m1}$ and $C = 0.585\beta_{m1}^2 = 0.585(R_1/l_{m1})^2$. It is easily seen that this is the maximum of perveance for a beam through a tube with radius R_1 and length $2l_{m1}$.

We may finally discuss the beam shape at large distances from z_m. Since $R'' \propto R^{-1}$, the convergence 2β varies very slowly with z. Over a region, in which $R_m \ll R \leq R_{max}$, the beam shape is approximately conical with cone angle $2\beta(R_{max})$. [Note that Eq. (26.10) gives $\beta \to \infty$ for $R \to \infty$.]

For given values of R_1 and β_1 at z_1, a conical beam shape is obtained for $C \to 0$, where $R_m = 0$ and $l_{m1} = l_1$. When R_m is negligibly small, it may be said that there is a beam crossover, but strictly speaking, a true crossover is avoided as long as the charge distribution is adequately described by a continuous density function.

So far, the ideal round beam in a drift region has been treated. With an outer electric field with axial symmetry, the ray equation is

$$r'' + (V'/2V)r' + (V''/4V)r - C_c/r = 0. \quad (26.11)$$

It may be noted that V is normally inserted as the potential in the absence of beam charge, and here it is assumed that charge distribution on electrodes is not affected by the presence of beam charge. Furthermore, it is assumed that the axial component of the space-charge field is small. It may, however, be mentioned that in a retardation system, in which particles are slowed down to extremely low velocities, these assumptions are not fulfilled; we shall exclude such cases from further consideration.

If there are several crossovers in an apparatus, it is seen that, for intensity increasing from zero, the space-charge effect must first be taken into account at the crossover characterized by having the lowest value $\beta^2 V^{3/2}$, since R_m depends very strongly on this quantity. (Note that $\beta^2 V^{3/2}$ is inversely proportional to $C\beta^{-2}$.) Normally, it is preferred that this crossover is the final focus so that the space-charge effect is small at intermediate crossovers. One reason is that with $l_{m1} \doteq l_1$ for intermediate crossovers, the focusing does not depend sensitively on the intensity; another reason is that for a nonideal beam, the beam quality may be reduced by space-charge effects (see Section 27).

It is an important feature of the round beam that the space-charge term $-C_c/r$ in Eq. (26.11) is small in regions where R is large.

Consider now the ideal flat beam in a drift region. By means of Gauss' law, it is found that E_y is given by

$$E_y = (y/Y)(I_1/2\varepsilon)(m/2eV)^{1/2},$$

where $2Y$ is the beam width, $|y| \leq Y$, and I_1 is the current per unit of beam height. It is easily shown that $y'' = Y''y/Y$, so that $y' \propto y$ for all z.

The differential equation for Y is found to be

$$Y'' = D_1 = \pi C_1, \tag{26.12}$$

where C_1 is the value of C obtained from Eq. (26.1) with I_1 inserted for I. Since $Y'' = \text{const}$, the space-charge effect is not limited to regions where Y is small.

In a region where Y does not change sign, the curve for Y versus z is a parabola. For given values of Y_1 and $\beta_1 = -Y_1'$ at z_1, a true crossover is obtained in the intensity range given by

$$D_1 < \beta_1^2/2Y_1, \tag{26.13}$$

while the crossover is avoided for higher intensities. Thus, with line focusing a true crossover may be obtained, while as seen above this is impossible with point focusing.

27. Focusing of a Nonideal Beam and Various Beam Effects

The effect of a finite beam emittance will be discussed for a round beam focused in a drift region, but before this, various other effects may be mentioned in passing, even though this review will actually be the major part of the section.

27. FOCUSING OF A NONIDEAL BEAM

With a nonuniform current density, the condition $r' \propto r$ cannot be fulfilled for all z. Consider for instance a hollow beam which at z_m has zero density for $r < R_0$ and a uniform density in the shell $R_0 \leqslant r \leqslant R_m$; all trajectories have $\varphi' = 0$ and $r' = 0$ at z_m. The inner beam surface is seen to be the cylinder with radius R_0, and the outer surface may be derived from $R'' = C/R$. The laminar structure is conserved, but $r' \propto r$ is fulfilled only for $z = z_m$. A lens at $z_L > z_m$ will therefore not give the proper focusing.

Consider next an intermediate focus formed by a lens with spherical aberration. Here, the current density is nonuniform, and furthermore, the beam structure may be nonlaminar. The space-charge effect in the focus region may then seriously reduce the beam quality. If possible, a design with no intermediate focus should be chosen, and if there has to be one, the aberrations and space-charge effects must be small. As pointed out in Section 26, the space-charge effect is small when $\beta^2 V^{3/2}$ is large.

Collisions with molecules of residual gas in the vacuum may give effects of importance which, even though they are not of a purely optical nature, will be mentioned here.

Small-angle scattering of ions limits the obtainable purity in isotope separation. The isotope to be collected may have a very low abundance, and due to scattering in collisions with molecules of residual gas, ions from the intense beams of other isotopes are collected as impurities. In some isotope separators, the separation is performed in two stages. Here, a diaphragm is placed at the focal plane for the first separation, and the beam transmitted to the second magnet has a small acceptance angle which, to a large extent, excludes the impurity ions.

Another effect of collisions with molecules of residual gas is that ions may have their charge state changed; for instance, singly charged ions may be neutralized, and since this may occur in very weak collisions, the neutrals are not scattered out of the beam. If the effect occurs in a retardation field, the neutrals in the transmitted beam have higher energies than the ions.

The charge-changing process is utilized in a tandem accelerator where negative ions are accelerated to a terminal with positive potential V_a; here the charge number is changed from -1 to $+n$ by penetration through a foil or a gas cell. In a second acceleration tube, the $n+$ ions are accelerated, and when both source and target are at ground potential,

7. SPACE CHARGE AND BEAM PRODUCTION

the resulting ion energy is $(1+n)V_a$. The stripping process in the V_a terminal may give various charge states, and a definite value of n, and thus of the energy, is selected by means of a deflection magnet.

Also ionization of the molecules of residual gas may be of importance both for electron beams and ion beams since it affects the space-charge effect. In the ionization process, fast electrons are ejected, and therefore a positive charge density due to residual-gas ions is present in the beam. (Most of the residual-gas ions have only thermal velocities.) For an electron beam through a gas cell, the charge density is mainly due to residual-gas ions, and the positive potential in the beam is proportional to the product of beam current and gas density. In Auger spectroscopy, where the energy of the ejected electrons is measured, the positive beam potential gives rise to a significant correction.

If a beam hits a diaphragm, a large number of secondary electrons is produced, and this has several consequences:

i. In an ion accelerator, electrons will be accelerated towards the source where X-rays are then produced. This may be prevented by means of a negative potential barrier, a transverse electric field, or a transverse magnetic field.

ii. Beam current is normally measured by collecting the charge in a Faraday cage connected to ground through an ammeter. Secondary electrons must be prevented from leaving the cage, and furthermore, the cage and the wire carrying the current from the cage must be shielded against secondary electrons produced elsewhere.

iii. Electrons may be reflected from the walls or produce new secondary electrons, and slow electrons tend to be present everywhere. This gives a noise problem, for instance in the Auger spectroscopy dealing with low-energy electrons. Careful screening is necessary, and internal surfaces in the spectrometer must have a low secondary emission yield and a low reflection yield; rough surfaces of graphite may be used.

iv. Charge buildup on insulators gives uncontrollable electric fields, and in accelerators, spectrometers, etc., the electrodes must provide an efficient screening of the beam region. Insulating surface layers may make charge buildup possible even on electrodes. A good shape of a diaphragm is shown in Fig. 27.1.

v. A charge density of slow electrons may be present in an ion beam and partly compensate for the charge of ions. This occurs only in regions

27. FOCUSING OF A NONIDEAL BEAM

FIGURE 27.1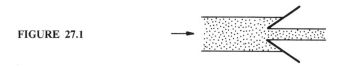

with no outer electric field. If the beam is screened against electric fields and focused by means of magnetic fields, full compensation may be obtained so that the space-charge effect is eliminated. The magnetic focusing, utilizing space-charge compensation, is applied for high beam currents and especially for flat beams, because a high current can best be obtained from a slit source, and the focusing is best performed as line focusing. An example is the *calutron*; this is an enormous 180° isotope separator which is able to handle beams of many amperes. It produces large quantities of isotopes with many valuable applications.

For a flat beam, point focusing may be obtained by means of a cylindrical lens with a large opening, but irregular structures may arise from the parts of the beam originating from the ends of the source slit.

A round beam from a small circular source is normally best suited for point focusing and for the production of a narrow pencil beam; good line focusing may also be obtained. With large beam radii in lens regions, the space-charge effect is small even with electrostatic lenses in which there is no space-charge compensation. At beam crossovers, it may be valuable to have space-charge compensation.

Consider now a round beam with a finite emittance. In a drift region with a constant potential V, it is nearly an ideal beam with uniform current density; there is no space-charge compensation and no effect of collisions with molecules of residual gas. The beam is specified by the perveance C and by R_1 and $\beta_1 = -R_1' > 0$ at z_1 (see Section 26), and furthermore by the normalized emittance τ (see Section 4).

The focus width for $\tau = 0$, $C \neq 0$, is $2R_m$, and the focus is located at $z_m = z_1 + l_{m1}$ (Section 26). For $C = 0$, $\tau \neq 0$, the focus width $2R_0$ may be derived from

$$\tau = \pi R_0^2 \pi \beta_1^2 2emV, \qquad (27.1)$$

and the focus is located at $z_0 = z_1 + l_1$ (Section 4). The intermediate case will not be treated in detail, but for a given value of C, an emittance τ_c

will be determined such that the space-charge effect dominates for $\tau < \tau_c$. Here, we may use the criterion that R_0 given by Eq. (27.1) for $\tau = \tau_c$ is equal to R_m given by Eq. (26.6), and τ_c is then found to be given by

$$\tau_c = 2emV(\pi R_1 \beta_1)^2 \exp(-\beta_1^2/C). \tag{27.2}$$

The conservation rule for emittance is valid also with space charge, and τ is then given by the source emittance,

$$\tau = \pi R_s^2 \pi 2emV_T, \tag{27.3}$$

where R_s is the radius of the circular source area and V_T represents the thermal energy of particles released from the source, $eV_T = kT$ (see Section 4). For $\tau = \tau_c$, it is found that $R_s = R_c$, where R_c is given by

$$R_c = R_1(V/V_T)^{1/2}\beta_1 \exp(-\beta_1^2/2C). \tag{27.4}$$

If $R_s < R_c$, the low-emittance beam with $\tau < \tau_c$ is produced.

The current density at the source is $i = I/\pi R_s^2$, and for a given type of source, there is a maximum i_{max} for the source emission. This gives the condition $R_s > R_{s\,min}$ given by

$$R_{s\,min} = (I/\pi i_{max})^{1/2}. \tag{27.5}$$

It is then seen that it is possible to produce the low-emittance beam when $R_{s\,min} < R_c$, which gives the condition

$$i_{max}/V_T > (I/\pi R_1^2 V)\beta_1^{-2} \exp(\beta_1^2/C). \tag{27.6}$$

The lower limit for i_{max}/V_T depends very strongly on $\beta_1^2 C^{-1}$.

A high-energy accelerator can normally not produce a low-emittance beam, and for $\tau > \tau_c$, the minimum radius of the focus is given by

$$R_{0\,min} = R_{s\,min}\beta_1^{-1}(V_T/V)^{1/2} = \beta_1^{-1}(V_T I/\pi i_{max} V)^{1/2}, \tag{27.7}$$

which is obtained by means of Eqs. (27.1), (27.3), and (27.5).

28. Electron Extraction

Electrons are released from a cathode by thermionic emission and may be accelerated toward the extraction electrode. The current density is limited by space charge, as will be seen below, but for strong extraction

28. ELECTRON EXTRACTION

fields, a saturation i_{max} is reached. The saturation is at least approximately given by Richardson's equation $i_{max} = AT^2 \exp(e\Phi/kT)$, where A depends on the surface and the work function Φ usually ranges from 2 to 6 V. It is seen that i_{max} depends strongly on the temperature. For oxide cathodes, about 0.5 A/cm^2 is obtained at 1160°K, corresponding to $V_T = 0.1$ V. In the following it is assumed that saturation does not occur.

The extraction system may be designed on the basis of relations for an ideal diode where the thermal energy is infinitely small.

In a plane diode, the cathode is the plane $z = 0$ and the anode is the plane $z = d$. The cathode potential is zero, and the anode has the potential V_a. Using e for the numerical charge of an electron, the kinetic energy is

$$\tfrac{1}{2}m\dot{z}^2 = eV, \qquad V = (z/d)V_a - V_q,$$

where $V_q(z)$ is due to space charge. It follows from Gauss's law that

$$V'' = -V_q'' = \varepsilon^{-1}\eta e,$$

where $\eta = \eta(z)$ is the electron density.

The current density i, which is the same for all z, is given by

$$i = \eta e\dot{z} = \eta e(2eV/m)^{1/2},$$

and V'' may then be expressed as

$$V'' = \varepsilon^{-1}i(m/2eV)^{1/2}. \tag{28.1}$$

The function

$$V = V_a(z/d)^{4/3} \tag{28.2}$$

may be seen to be the solution to Eq. (28.1). The field strength is then numerically given by

$$E = \tfrac{4}{3}(V_a/d)(z/d)^{1/3}. \tag{28.3}$$

It is noted that $E = 0$ at the cathode, while $V'' = \infty$ corresponding to infinite charge density at the cathode.

The current density is derived by inserting V and V'' in Eq. (28.1),

$$i_0 = \tfrac{4}{9}\sqrt{2}\,\varepsilon(e^{1/2}V_a^{3/2}/m^{1/2}d^2). \tag{28.4}$$

The index 0 is, in the following, associated with the case of a plane diode.

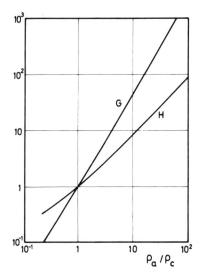

FIGURE 28.1†

In a spherical diode, the cathode is a sphere with radius ρ_c, and the anode is a concentric sphere with radius ρ_a. For $\rho_a > \rho_c$, an outgoing radial flow is obtained, and for $\rho_a < \rho_c$ an incoming flow. The current density i at the cathode may be expressed as

$$i = Gi_0, \tag{28.5}$$

where i_0 is given by Eq. (28.4) for $d = |\rho_a - \rho_c|$, and G is a function of ρ_a/ρ_c. This function is shown in Fig. 28.1. In the double logarithmic plot, the curve is approximately a straight line, giving

$$G = (\rho_a/\rho_c)^{1.63}. \tag{28.6}$$

Consider cases for which d has a fixed value. For $\rho_c < \rho_a = \rho_c + d$, the current $4\pi\rho_c^2 Gi_0$ is given by $4\pi i_0 \rho_c^{0.37}(\rho_c + d)^{1.63}$, which approaches zero for $\rho_c \to 0$, while the current density Gi_0 at the cathode increases towards infinity. For $\rho_c \ll d$, the space-charge field is only significant in a small region at the cathode. For $\rho_a < \rho_c = \rho_a + d$, the current is given by $4\pi i_0(\rho_a + d)^{0.37}\rho_a^{1.63}$, which approaches zero for $\rho_a \to 0$.

In a cylindrical diode, the cathode and anode are coaxial cylinders

† I. Chavet and R. Bernas, *Nucl. Instrum. Methods* **47**, 77 (1967).

28. ELECTRON EXTRACTION

with radii ρ_c and ρ_a, respectively. The current density at the cathode is given by

$$i = Hi_0, \qquad (28.7)$$

where H is a function of ρ_a/ρ_c as shown in Fig. 28.1. A good approximation to H is obtained as

$$H \doteq 1 + 4(\rho_a - \rho_c)/5\rho_c. \qquad (28.8)$$

It is seen that $H \doteq 4d/5\rho_c$ for $\rho_c \ll d$; the current $2\pi\rho_c Hi_0$ per unit length is then independent of ρ_c, which shows that a finite current is obtained for $\rho_c = 0$. For $\rho_a \ll d$, one gets $H \doteq 0.2$, and a finite current is obtained for $\rho_a = 0$.

Both for the spherical and the cylindrical diode one obtains the current density i_0, when $\rho_c \gg d = |\rho_a - \rho_c|$.

In a triode, a grid with potential V_g is placed between the cathode and the anode. Its distance d_g from the cathode is small compared to the distance d between cathode and anode. Normally, V_g is slightly negative so that the electron current to the grid is zero. The current to the anode is limited by space charge between cathode and grid, and the voltage across this diode region is $aV_a + V_g$, where the term aV_a is due to the openings in the grid, and one has $a \ll 1$. For a plane triode, the current density is given by

$$i_0 = \tfrac{4}{9}\sqrt{2}\varepsilon(e/m)^{1/2}(aV_a + V_g)^{3/2}d_g^{-2}.$$

It is noted that the current to the anode is controlled by the grid potential and that no current is drawn from the voltage generator V_g.

We shall now consider an extraction system. A positive extraction electrode with an aperture, through which the beam may pass, is placed in front of the cathode. An ideal flat beam may be obtained from a cylindrical diode with a narrow slit in the anode, and similarly, an ideal round beam may be obtained from a spherical diode with a small circular aperture in the anode. The aperture gives a lens action, and for the round beam, the focal length f of the thin lens is given by $f = 4V_a/\Delta V'$. The value of V' at $z = d$ in the diode region is derived from the potential function $V(z)$ for the diode. For a parallel beam extracted from a plane diode through a circular aperture to a drift region V_a, one finds $f = -3d$.

Frequently, it is required that no current flows to the extraction

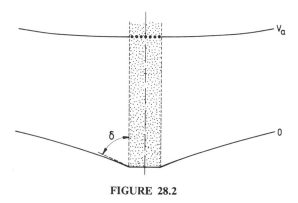

FIGURE 28.2

electrode, i.e., that the total beam should pass through the extraction aperture. We may, however, first discuss the case illustrated in Fig. 28.2, where a grid across the extraction aperture defines a diode region for the beam of finite width. Such designs were first investigated by J. R. Pierce.

Let it be required, for example, that the beam in the diode region should be a flat, parallel beam. The distribution of space charge in this beam is known, and the potential in the beam and on its surface, including the end surfaces, is given by $V = V_a(z/d)^{4/3}$. Introducing the known charge distribution in an empty space, a field is produced in the charge-free space outside the beam, and here the potential is the solution to the Laplace equation $\nabla^2 V = 0$ with the boundary condition given by the potential $V = V_a(z/d)^{4/3}$ on the beam surface. When the field is calculated, one electrode may be shaped in accordance with the equipotential surface for $V = 0$, and another as the V_a potential surface. This gives a geometry for which the ideal beam of finite width can be produced in the diode region.

The Pierce geometry for $d/2Y_s = 4$ is shown in Fig. 28.2. The angle δ between the beam surface and the electrode on zero potential is known as the Pierce angle which, for all beam shapes, has the value 67.5°.

When the grid across the extraction aperture is removed, the optimum geometry for a narrow beam is approximately the Pierce geometry, and the lens action of the aperture can easily be evaluated. Thus, if the Pierce geometry is known, it is not difficult to treat the case of a narrow beam. In other cases, however, it is very complicated to design an extraction

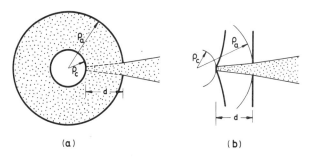

(a)   (b)

FIGURE 28.3

system giving the nearly ideal beam. (A complicated case, which has been extensively investigated, is the design of a high perveance electron gun; a concave cathode which is almost a semi-sphere is used, and the beam is compressed to a small diameter through the extraction aperture so that a narrow and very intense beam is obtained.)

The relations for a narrow, round beam may now be derived. The beam is obtained either from a spherical diode with a small aperture in the anode (Fig. 28.3a), or from an extraction system with Pierce geometry (Fig. 28.3b). The beam divergence in the diode region is 2α where $\alpha = R_s/\rho_c$, and it is seen that $\rho_a/\rho_c = 1 + \alpha d/R_s$. A concave cathode gives $\alpha < 0$, and here one must have $\rho_c > d$. The current I is given by

$$I = GI_0, \qquad I_0 = i_0 \pi R_s^2, \tag{28.9}$$

where according to Eq. (28.6),

$$G = (1 + \alpha d/R_s)^{1.63}. \tag{28.10}$$

By means of Eqs. (26.1) and (28.4), it is found that the normalized perveance C_a at the extraction aperture is given by

$$C_a = (G/9)(R_s/d)^2, \tag{28.11}$$

and it is noted that C_a is fully determined by α and d/R_s.

The beam leaving the extraction system is characterized by C_a, the beam radius $R_a = R_s + \alpha d$, and the divergence $2\alpha_a$, where $\alpha_a = R'$ outside the aperture. For simplicity, we may use the approximation $\alpha_a \doteq \alpha$, in which the lens action is not taken into account. According to Section 26, the beam shape in a drift region outside the extraction electrode depends strongly on $C_a \alpha_a^{-2}$.

7. SPACE CHARGE AND BEAM PRODUCITON

Two extreme cases may be considered, namely the case with a small cathode,

$$R_s \ll \alpha d, \quad C_a \doteq (\alpha^{1.63}/9)(R_s/d)^{0.37}, \quad C_a \alpha_a^{-2} \doteq \tfrac{1}{9}(R_s/\alpha d)^{0.37}, \tag{28.12}$$

and the case with a large cathode

$$R_s \gg \alpha d, \quad C_a \doteq \tfrac{1}{9}(R_s/d)^2, \quad C_a \alpha_a^{-2} \doteq \tfrac{1}{9}(R_s/\alpha d)^2. \tag{28.13}$$

With a small cathode the perveance is small, and a nearly conical beam shape is obtained outside the extraction aperture.

With a large cathode, a high perveance is obtained, and the space-charge effect is large in the extracted beam. This beam may be characterized by C_a and by a minimum radius R_m; if $\alpha_a > 0$, R_m is the minimum radius of the beam extrapolated behind the extraction electrode. For $R_s \gg \alpha d$, one has $R_m \approx R_s$.

It should be noted that there is an absolute limitation on the perveance C_a. The simple expression, Eq. (28.13), may be used for $2R_s \lesssim 0.2d$, which gives $C_a \lesssim 0.001$, but due to complicated effects of the extraction aperture, it is difficult to obtain a good beam quality when d/R_s is small. With an advanced design with a large, concave cathode, a perveance of about 0.1 may be reached.

When the extracted beam is projected into a drift region with potential V by a lens system, the perveance C of the final beam is given by

$$C = C_a(V_a/V)^{3/2}. \tag{28.14}$$

If the type $(\alpha, d/R_s)$ of the extraction system is chosen so that C_a has a given value, and if C is specified, then the extraction voltage V_a is given by

$$V_a = V(C/C_a)^{2/3}. \tag{28.15}$$

If, furthermore, it is specified that the emittance should be low, $\tau < \tau_c$, so that it is the space-charge effect which limits the focusing, the size of the extraction system is limited by the condition $R_s < R_c$, where R_c is given by Eq. (27.4). The beam with $\tau < \tau_c$ can be produced when $R_c < R_{s\,min} = (I/\pi i_{max})$ [Eq. (27.5)]. When C is higher than the highest obtainable C_a, it is seen that a design with a deceleration lens must be used so that the beam with a high perveance is produced by acceleration and deceleration.

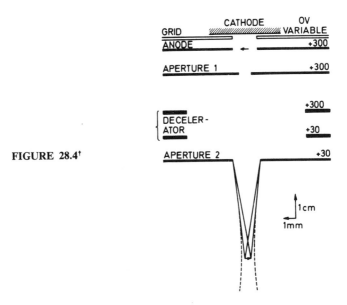

FIGURE 28.4[†]

A design of an electron gun of this type $(C > C_a, \tau < \tau_c)$ is shown in Fig. 28.4. Here the extraction system has the structure of a plane triode, and the current is controlled by the grid voltage.

Finally the flat beam may be considered. Here the relations Eqs. (28.7) and (28.8) for the cylindrical diode may be used. For $\rho_c \ll d$, one has $I_1 = 2\alpha \rho_c H i_0 \doteq 2\alpha \cdot 0.8 i_0 d$, while I_1 for $\rho_a \ll d$ is given by $I_1 \doteq |2\alpha| 0.2 i_0 d$. Finite currents are obtained both for a beam which diverges from an infinitely narrow cathode and for a beam converging towards a very narrow extraction aperture.

For a concave cathode with $\rho_c < d$, the beam has a crossover at $z = \rho_c$ in the extraction gap, and the diode representation can be used for the region $0 \leqslant z \leqslant \rho_c$.

29. Ion Extraction

Ions may be produced in the plasma of a discharge in a gas at low pressure, and the ionization is here mainly due to electron-atom collisions and electron-molecule collisions. A beam of ions may be extracted from the discharge chamber through a small outlet opening.

[†] J. A. Simpson and C. E. Kuyatt, *Rev. Sci. Instrum.* **34**, 265 (1963).

There are many types of ion sources, but it is beyond the scope of this book to review this large and active field of research. Only a few general features of relevance to ion extraction may be mentioned.

For several elements in the periodic table, ions are formed directly from gases; for example, the ions N^+, N^{2+}, N_2^+ are produced in a discharge in N_2. Other elements are obtained from solid charge materials heated in an oven so that a suitable vapour pressure is obtained. For example K^+ ions can be produced in this way from KCl. In such cases, the temperature of the discharge chamber must be high so that a loss of charge material due to condensation on its walls is avoided. In some cases, a gaseous compound may be formed by chemical reactions; for example W^+ ions can be obtained from $WO_3 + CCl_4$, which gives the gaseous compound. By various techniques it is possible to produce ions of nearly all elements in the periodic table. Normally the extracted beam is a mixture of ions from several different elements, and therefore an ion accelerator consists of the ion source, the acceleration system, and a deflection magnet. When the magnet gives a large mass dispersion, isotopically pure beams are produced.

As mentioned above, charge material may be lost on the walls of the discharge chamber. Material is also lost by the flow through the outlet of nonionized atoms and molecules; here it is an important factor that a discharge can only be operated at pressures above a certain minimum, depending on the gas and the type of ion source, and the pressure should be kept close to this minimum. Furthermore, the degree of ionization should be high and the source outlet small. (A low minimum pressure is obtained by introducing a filament emitting electrons, and by introducing a magnetic field reducing the loss of electrons to side walls.)

The perveance of the extraction system is limited, and here it is noted that the perveance with different ions in the beam is derived from $\sum I_{M,n}(M/n)^{1/2}$. Therefore, a high abundance of the wanted type of ions is desirable; this abundance depends on the cleanliness of the chamber and on the parameters controlling the discharge.

The following features are of particular relevance for ion extraction.

The plasma is an equipotential region with equal densities of negative and positive charge. Its potential is close to (slightly lower than) the potential of the anode in the discharge chamber. When the plasma is facing an electrode with a lower potential, a quite well defined boundary surface is formed, and the current of positive ions to the electrode is

29. ION EXTRACTION

FIGURE 29.1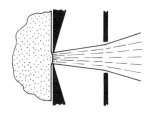

space-charge limited in the same way as the current in a diode. On the plasma boundary, the field strength is zero. The thickness of the diode layer increases with the potential difference.

An extraction system is shown in Fig. 29.1. From the plasma boundary across the outlet, the ions are extracted towards the extraction electrode which has the potential $-U$ with respect to the plasma potential. The shape of the plasma surface is determined by the condition that the space-charge limited current density outside the surface is determined by the product of ion density and ion velocity in the plasma. In the figure the plasma boundary is convex which, according to Section 28, corresponds to a high current density. If the plasma density and thus the current density is reduced, the boundary surface becomes concave. It is seen from Fig. 28.1 that rather small changes of the shape of the boundary surface correspond to large variations of the current density. However, under extreme conditions, the plasma boundary may become considerably deformed (Fig. 29.2), but such cases will not be considered here.

FIGURE 29.2

It is noted that the ion-source parameters, i.e., pressure, filament current, etc., affect the boundary curvature only through the resulting current density i. If two parameters are varied in such a way that i is kept constant, then the curvature of the boundary is constant. Thus, the cumulative parameter i can be considered as one of the extraction parameters.

Assume now that the outlet electrode and the extraction electrode are shaped in accordance with the Pierce geometry for a round, narrow beam with given values of R_s, d, and α (Section 28). Provided the current density is properly adjusted, this beam is produced, and all the relations derived in Section 28 for the round beam are valid. If the current density is not correct, the beam shape in the diode region is not conical, and the plasma boundary is not spherical; the relations obtained from the spherical diode are then not valid, and the beam quality is reduced.

Nevertheless, as established by Chavet and Bernas,[†] the representation by simple diode structures is approximately correct over the useful range of divergence for a beam from a very small outlet. Even though the beam shape is not exactly conical, the angle α at the source may be approximated as divergence at some distance from the outlet. The field distribution near the source is clearly dominated by the space charge field, since a very high current density is obtained in this region.

When the subsequent focusing system projects the beam as a convergent beam, and a small focus width is essential, there are several advantages obtained by extracting a divergent beam from a very small source outlet, $R_s \ll \alpha d$. The advantages are the following: (i) The beam quality is good, when the emittance is small. (ii) With a high current density at the outlet, one obtains a high ratio between the current and the loss of neutrals. (iii) The space-charge effect is small outside the extraction electrode.

On the other hand, the perveance is low, which means that a high extraction voltage must be applied for obtaining a given current. It is also noted that the divergent mode can only be used when a sufficiently dense plasma can be produced.

Chavet and Bernas performed an experimental investigation for the case of a flat beam, and they verified that with Pierce geometry for $\alpha = 0$, the cylindrical-diode representation could be applied for divergent

---

† I. Chavet and R. Bernas, *Nucl. Instrum. Methods* **47**, 77 (1967).

29. ION EXTRACTION

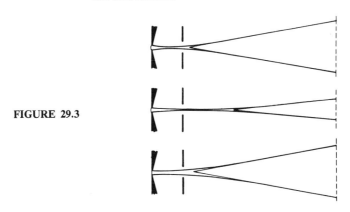

FIGURE 29.3

beams and for nearly parallel beams, $R_s \gtrsim \alpha d$. For instance the model gives an approximately correct value of $dI/d\alpha$ for $\alpha = 0$.

The case of a round beam was not investigated experimentally. Since the outer field has a stronger influence in this case, it may be expected that $dI/d\alpha$ for $\alpha = 0$ is smaller than predicted by the model. This implies that α may depend sensitively on the plasma density, which determines I.

In Fig. 29.3 it is illustrated that for $\alpha \approx 0$, small variations of α may give large variations of the position of the virtual source point to be imaged, and fluctuations of the plasma density may therefore reduce the beam quality. When R_s/d is not very small, the divergent mode of extraction can not be used, and stable running conditions are best obtained with a concave plasma boundary, which is obtained for a reduced current. It is thus seen that fluctuations of plasma density make it difficult to utilize a high perveance; other difficulties have been mentioned above.

It is concluded that for a specified current, the source outlet should be as small as possible, while the extraction length d should be large compared to R_s/α, which implies that the extraction voltage must be high. This conclusion applies when it is essential to obtain a small width of the final focus.

A different situation arises when a pencil beam should be produced, i.e., a narrow and nearly parallel beam. Here, a high brightness at $(x, x', y, y') = (0, 0, 0, 0)$ is essential for the final beam. This can best be obtained with a parallel or slightly convergent beam in the extraction region, and a narrow beam all the way throughout the acceleration system.

Problems

CHAPTER 2

Problems 2.1–2.5 deal with image formation in one projection.

2.1 Lens 1 for which $f_{i1} = f_{s1} = f_1$, $H_{i1} = H_{s1} = H_1$, is combined with lens 2 for which $f_{i2} = f_{s2} = f_2$, $H_{i2} = H_{s2} = H_2$, so that $H_2 = H_1 + l$.

Prove that $d_{12} = 1$. Determine $[H_2 \leftarrow H_1]$, $1/f$, H_i, and H_s. Give the results for $f_2 = f_1$ and for $f_2 = -f_1$.

2.2 A system consists of two lenses for which $f_{s1} = f_{i2}$, $f_{i1} = f_{s2}$, $H_{i1} = H_{s1} = H_1$, and $H_{i2} = H_{s2} = H_2 = H_1 + l$.

Prove that $d_{12} = 1$. Determine $[H_2 \leftarrow H_1]$, f, H_i and H_s.

2.3 In lens 1 particles are accelerated from energy T_0 to $T_1 = 2T_0$, and in lens 2 from T_1 to $T_2 = 3T_0$.

a. Determine f_{i1}/f_{s1} and f_{i2}/f_{s2}.
b. Assume now $f_{s2} = f_{i1}$, and let a source be placed at F_{s1}. Determine image position and magnification.

2.4 In a lens particles are accelerated from T_1 to $T_2 = cT_1$. A source is placed at $F_s - p$. Assume that f_s and f_i are known.

132

a. Determine image position and magnification.
b. Give the results for $p = f_s$.

2.5 Determine $[K_i \leftarrow K_s] = [K_i \leftarrow F_i][F_i \leftarrow F_s][F_s \leftarrow K_s]$ where $K_i = F_i - f_s$ and $K_s = F_s + f_i$.

Prove that

$$\begin{bmatrix} 0 \\ \alpha \end{bmatrix} = [K_i \leftarrow K_s] \begin{bmatrix} 0 \\ \alpha \end{bmatrix}.$$

A source at $K_s - l_1$ is imaged at $K_i + l_2$; find l_2 and the magnification. With a slit-diaphragm w at K_s the beam divergence at a source point is $\theta_1 = w/l_1$; find the convergence $-\theta_2$ at the image.

2.6 In a field-free region diaphragms are inserted at z_1 and $z_2 = z_1 + l$.

a. Find the acceptance a_y for a slit–slit collimator, $|y_1| \leq \tfrac{1}{2}w_1$, $|y_2| \leq \tfrac{1}{2}w_2$.
b. Find a_{xy} for two rectangular slits, w_{x1}, w_{y1}, and w_{x2}, w_{y2}.
c. Find a_{xy} for a hole–hole collimator, $r_1 = (x_1^2 + y_1^2)^{1/2} \leq R_1$, $r_2 \leq R_2$.

2.7 A beam is collimated so that $\omega_{xy} = a_{xy}$ in a region, where the energy is T_1. Find the emittance ω_{xy} in another region, where the energy is $T_2 = cT_1$.

2.8 By means of long slits, x collimation giving $\omega_x = a_x$ is performed in region 1, where the energy is T_1, while y collimation $\omega_y = a_y$ is performed in region 2, where the energy is $T_2 = cT_1$.

Determine the emittance ω_{xy} in region 1 and in region 2.

2.9 From a small circular source area with radius R_s particles are emitted with momentum p_s, and a hole-diaphragm R is placed at distance l from the source.

a. Find the geometrical emittance ω_{xy} at the source, and find the normalized emittance τ_{xy}.

b. It is now required that the beam in a subsequent section of apparatus is transmitted through a hole–hole collimator with acceptance a_{xy}, which is smaller than the emittance ω_{xy} at the source. What is the necessary acceleration, $c = T/T_s$, of the intermediate section of apparatus?

Chapter 3

3.1 Consider two similar systems with electric and magnetic fields. In system e the particles are electrons, and in i we have (M, n) ions. Determine scaling factors from e to i for linear dimension, energy, magnetic and electric field strength.

3.2 Ions of Ar^{40} may have $n = 1, 2, 3, \ldots$. Calculate for each n the magnetic field B giving $R = 60$ cm,

 i. when $T = 100$ keV, and
 ii. when $T = neV_{acc}$ where $V_{acc} = 100$ kV.

3.3 For electrons with energy $T = (\mu - 1)mc^2$ moving in a magnetic field one has $BR = K\sqrt{T}$, where $K = K_0$ for $\mu - 1 \ll 1$.

Determine K/K_0 as function of $\mu - 1$.

3.4 For electrons with energy $T = (\mu - 1)mc^2$ moving in an electric field one has $ER = KT$, where $K = K_0$ for $\mu - 1 \ll 1$.

Determine K/K_0 as function of $\mu - 1$.

3.5 According to Eq. (15.10) the focal length of a weak magnetic lens is given by

$$1/f = (e^2/8Tm) \int B_z^2 \, dz, \quad \text{mks units.}$$

Determine for electrons and for (M, n) ions the constant K in the expression

$$1/f = (K/T) \int B_z^2 \, dz, \quad \text{cm, keV, G.}$$

Chapter 4

4.1 Concentric, spherical electrodes with radii r_1 and r_2 have potentials $-V/2$ and $+V/2$, respectively.

a. Prove that the radial field E is proportional to r^{-2}, and determine the potential function $V(r)$.
b. Find the value R of r for which $V(r) = 0$, and determine $E(R)$.

4.2 Concentric, cylindrical electrodes with radii r_1 and r_2 have potentials $-V/2$ and $+V/2$, respectively.

Prove that $E \propto r^{-1}$, and determine $V(r)$. Find the value R of r for which $V(r) = 0$, and determine $E(R)$.

4.3 The north and south pole faces of a magnet are two half-planes both bounded by the z axis of an $zr\varphi$ frame. Prove that $B_z = B_r = 0$ and $B_\varphi \propto r^{-1}$ (Fig. 24.1).

4.4 Prove that the field $B(1/2)$ is obtained for a paraxial region in a magnet with conical pole faces, when the pole distance is given by $2h = (1+r/R)h_0 \ll R$ (Fig. 19.1).

4.5 Let the shielding geometry, Fig. 11.3, be given by $d = 2b = 2s$ and $t = \infty$.

a. Determine the effective field boundary z_{eff} by means of Fig. 11.4.
b. Make a field-line picture by means of the square cell rule; derive $E_y(z)$ and determine z_{eff}.

4.6 Verify that the electric quadrupole field, Fig. 11.12, is given by Eq. (11.11), i.e., that $V(x,y)$ obtained from Eq. (11.11) is the solution to $\nabla^2 V = 0$.

Chapter 5

5.1 A plane electrode with a small, circular aperture is placed at $z = 0$. Determine the focal length f of the aperture lens, when the kinetic energy T as function of z is given by:

i. $T = T_0$ for $z < 0$, $T = T_0(1+z/l_0)$ for $z > 0$; l_0 is positive.
ii. $T = T_0(1-z/l_0)$ for $z < 0$, $T = T_0$ for $z > 0$.
iii. $T = T_0(1+z/l_0)$ for $z < 0$, $T = T_0$ for $z > 0$.

5.2 A plane electrode at $z = 0$ has a narrow slit-aperture, $|y| < w$. Determine f_y and f_x, when $T = T_0$ for $z < 0$, and $T = T_0(1+z/l_0)$ for $z > 0$.

5.3 A retardation system between $z_1 = -L$ and $z_2 = 0$ has a homogeneous field, and fine grids are inserted across apertures. The energy is T_0 for $z < -L$, and pT_0 for $z > 0$; here one has $p < 1$. At $z_1 = -L$ the beam has an ideal structure with $\varphi_1' = 0$ and $r_1' = -r_1/l_1$, where l_1 is the same for all trajectories.

a. Determine l_1 as function of p such as to obtain a focus at a given position $z = l_2$.

b. Consider next facet lenses of the grid at $z_2 = 0$, Fig. 12.2. The openings are $d \times d$ squares, but the lens action is approximately the same as for circular apertures with diameter d. Determine d such that a given small diameter D_0 of the beam is obtained at the focus, $z = l_2$.

c. Calculate l_1 and d for $L = 4$ cm, $p = 0.1$, $l_2 = 16$ cm, $D_0 = 0.1$ cm.

5.4 The electrodes of an unipotential lens (V, pV, V) are planes with small, circular apertures, placed at $z = -L, 0, L$. Outside the lens the energy is $T = -eV$.

a. Determine focusing matrix, principal planes, and focal points.

b. Find the source position $z = -l$ for which the image is obtained at $z = l$. Determine the beam radius R_0 at $z = 0$ for a beam with radius R_1 at $z = \pm L$ obtained from a source point at $z = -l$.

c. Calculate l/L and R_0/R_1 for $p = 4$, and for $p = 0.25$.

5.5 Consider an acceleration system of the type treated in Section 13, for which $L = 1$ unit of length, $z_1 = -0.5$ and $z_2 = 0.5$.

Calculate and plot f_1, F_1, f_2, F_2 as functions of p. (The figure will be similar to Fig. 14.2, which applies for an immersion lens.)

PROBLEMS 137

5.6 An immersion lens (V, U) of the type shown in Fig. 14.1a has electrode diameter $D = 5$ cm, and U is given by $U = 0.8\ V$. An approximation for f_2 is given by Eq. (12.6), and in this expression the integrand $(V'/V)^2$ may be obtained by means of Fig. 14.1b.

Determine f_2 and f_1.

5.7 In a quadrupole doublet the quadrupole lenses of length L are separated by a distance d. Suppose that Φ is small.

   a. Determine the focusing matrices.
   b. Determine for $d = L$ the source position for which exact stigmatic focusing is obtained, and calculate the magnifications m_x and m_y.

Chapter 6

6.1 Consider sector analyzers with $\Phi = \pi/2\varepsilon_y$; consider the yz projection.

   a. Prove that the focal points are situated on the sector boundaries.
   b. Prove that $l_1 l_{2y} = (R/\varepsilon_y)^2$, and that $m_y = \varepsilon_y l_{2y}/R = R/\varepsilon_y l_1$.
   c. Determine the coefficient D of energy dispersion as a function of l_1; use Eq. (6.4) with $y_\gamma = \gamma \varepsilon_y^{-2} R/\varkappa$, where $\varkappa = 1$ for $E(n)$, $\varkappa = 2$ for $B(n)$.
   d. Suppose $y_1' = y_1/l_1$ and $\gamma = 0$, and suppose that $|y| \leqslant \tfrac{1}{2}d$ inside the sector. Determine then $\theta_y = 2y'_{1\,\text{max}}$ as a function of l_1.
   e. Prove that $\omega_y \mathscr{R} = \theta_y D/m_y$, when $\mathscr{R}$ is given by Eq. (6.11). Determine this product as a function of l_1, $\omega_y \mathscr{R} = f(l_1)$.
   f. Determine Φ, $l_{2y}(l_1)$, and $f(l_1)$, for sector type $E(1)$, $E(2)$, $B(0)$, $B(1/2)$.

6.2 Special sector analyzers with $\Phi = \pi/2\varepsilon_y$; xz projection.

   i. $B(1/2)$ magnet with pole distance $2h = (1 + r/R)h_0$. For the paraxial region one has $|x| \leqslant h_0$. Suppose that $x_1' = x_1/l_1$, and determine $\theta_x = 2x'_{1\,\text{max}}$.

   ii. $B(0)$ magnet with pole distance $2h_0$. Determine θ_x.
   iii. $E(1)$ analyzer with a detector of height $2h_0$. Determine θ_x.

6.3 $B(1/2)$, $B(0)$, $E(1)$ analyzers with $\Phi = \pi/2\varepsilon_y$. For a source of height Δx_0 one has $\omega_x = \Delta x_0 \theta_x$, which gives

$$\omega \mathcal{R} = \omega_x \omega_y \mathcal{R} = \Delta x_0 g(l_1) \quad \text{where} \quad g(l_1) = \theta_x(l_1) f(l_1).$$

a. Determine $g(l_1)$ for the three types of analyzers.
b. Determine $g(l)$ where $l = R/\varepsilon_y$.

6.4 A spherical sector analyzer, $E(2)$, has $\Phi = 90°$ and $l_1 = l_2 = R$. Source and detector are small circular discs with radius r_0 normal to the spectrometer axis; the entire cone shell of transmission is utilized.

a. Prove that $r_0 \mathcal{R} = 2\sqrt{2} R$, and that $\omega = \pi r_0^2 \Omega$ where $\Omega = \sqrt{2\pi \theta_y}$.
b. Use θ_y as derived in Problem 6.1, and express the product $\omega \mathcal{R}$ in terms of r_0, d, R.

6.5 A cylindrical analyzer, $E(1)$, has $\Phi = \pi/\varepsilon_y = 127.2°$ and $l_1 = l_{2y} = 0$. The electrodes have potentials $\pm \frac{1}{2}V$ and radii $R \pm \frac{1}{2}d$. Source and detector are centered on the circle with radius R. The energy T of detected particles is given by $T = C_0 V$ where C_0 is given by $C_0 = eR/2d$, provided $d \ll R$.

Suppose now that the above design is applied also for d not very small compared to R. The ratio T/V will then deviate from C_0.

Determine the correct factor C of energy calibration, $T = CV$.

6.6 Consider an $E(1)$ analyzer with $l_1 \approx l_{2y} \approx l > 0$, and suppose that the sector boundaries have not been evaluated correctly. The angular error δ is the same at both boundaries.

Source and detector are mounted in accordance with an assumed geometry with $\Phi = \pi/2\varepsilon_y = 63.6°$ and $l_1 = l_{2y} = R/\sqrt{2}$, while the effective sector angle is given by $\Phi_{\text{eff}} = 63.6° + 2\delta$.

Determine $C = T/V$ and the ratio C/C_0 where C_0 is the factor of energy calibration for $\delta = 0$.

6.7 Consider the Browne–Buechner spectrograph, Fig. 20.1.

a. Prove that the focal curve is a hyperbola, and determine the curve.
b. Display the result in a figure with an energy scale indicated on the focal curve; use the energy for 90° deflection as unit.

c. Prove that second order focusing is obtained for 90° deflection.

d. A quadrupole lens with $f_x = -f_y = 2R$ is placed at the magnet entrance, and the source is then placed at a distance of $2R$ from the entrance; a diaphragm at the entrance has radius h_0, and the source radius is r_0. Determine then $\omega = \pi r_0^2 \Omega$ and $\omega \mathcal{R}$, where $\mathcal{R}$ is the resolving power for 90° deflection.

6.8 A magnet with homogeneous field and inclined sector boundaries has $\Phi = 90°$ and $\varepsilon_1 = \varepsilon_2 = \varepsilon = 26.6°$ ($\tan \varepsilon = 0.5$).

Find $\omega \mathcal{R}$ for a source of height Δx_0 placed at $l_1 = 2R$, when $|y| \leq \frac{1}{2}d$ and $|x| \leq h_0$ in the magnet; see Problems 6.1, 6.2, 6.3.

6.9 The spectrograph shown in Fig. 20.2 has $\Omega = 180°$, $\varepsilon_1 = \varepsilon_2 = -35.3°$ ($\tan 35.3° = 1/\sqrt{2}$).

a. Determine the focal curve, and prove that second order focusing is obtained for all points on the curve (see Section 21).

b. Suppose that the sign of ε_1 is changed, $\varepsilon_1 = +35.3°$, and investigate the x and y focusing which is then obtained.

CHAPTER 7

7.1 An ideal round beam in a drift region is specified by

$$R_1 = 0.5 \text{ cm}, \quad l_{m1} = 10 \text{ cm}, \quad R_m = 0.105 R_1.$$

Determine first l_1 and calculate then the current I for

i. electrons with energy 100 eV.
ii. protons with energy 1 keV.
iii. Ar^+ ions with energy 10 keV.

7.2 Find the maximum of beam current through a 100 cm long tube with diameter 5 cm for a beam of Ar^+ ions with energy 50 keV.

7.3 Consider for ideal round beams in a drift region the set of beam shapes for which given values are obtained for the radii R_0 at z_0 and R_1 at z_1. Here $z_1 - z_0$ is large compared to beam radii. One particular beam shape corresponds to maximum of perveance.

Prove that for $R_0 \ll R_1$ maximum perveance is obtained with a parallel beam at z_0, i.e., for beam shape specified by R_1, $R_m = R_0$, $l_{m1} = z_1 - z_0$.

7.4 An ion source has an outlet with radius R_s. The extraction length is $d \gg R_s$, and outside the extraction electrode there is a drift region of length $L \gg d$. The beam is collimated by a diaphragm with radius $R_1 \gg R_s$ placed at the end of the drift region.

Determine the maximum perveance for the collimated beam.

7.5 In drift region 1 with energy T_1 a beam shape is specified by R_1, l_{m1}, and $R_{m1}/R_1 = K_1 \ll 1$.

The beam is transmitted to drift region 2 with energy $T_2 = pT_1$, and focusing is adjusted such that given values are obtained for R_2 and l_{m2}.

Determine $R_{m2}/R_2 = K_2$ and find the value of p for which one has $K_2 = K_1$.

7.6 A pure beam of Ar^+ ions are extracted from an ion source with $R_s = 0.05$ cm: The extraction length is $d = 2$ cm, and the parallel mode of extraction is applied, $\alpha = 0$.

a. Determine the beam current I as a function of the voltage V_a.
b. Determine the emittance τ for $V_T = 0.1$ V.

7.7 Source and extraction system are the same as in Problem 7.6. The Ar^+ beam is focused in a drift region with potential $-V$ relative to the source. Here the beam is still a round beam, and the focusing is adjusted such that one has $R_1 = 0.5$ cm and $\beta_1 = 0.02$ rad.

a. For a low current, a radius R_0 of the focus is obtained; this focus width is due to the source emittance τ. Determine R_0 as a function of V.
b. Determine the perveance $C(V)$ such as to obtain $R_m = R_0$, where the minimum radius R_m is evaluated for an ideal beam, (Section 26).
c. Derive the current $I(V)$ from $C(V)$.
d. Derive the extraction voltage $V_a(V)$ from $C(V)$.
e. Find the value of V for which $V_a(V) = V$.

7.8 A flat Ar^+ beam is extracted from a source with a slit of width $2Y_s$ where $Y_s = 0.05$ cm. The extraction length is $d = 2$ cm, and the parallel mode of extraction is used, $\alpha = 0$.

  a. Determine the current I_1 per unit of beam height as a function of V_a.

  b. Determine the emittance τ_y for $V_T = 0.1$ V.

7.9 The flat Ar^+ beam from Problem 7.8 is focused in a drift region with potential $-V$, where one has $Y_1 = 0.5$ cm and $\beta_1 = 0.02$ rad.

  a. Determine, as a function of V, the focus width Y_0 due to emittance τ_y.

  b. Determine, as a function of V, the maximum of current I_1 for which a true crossover is obtained; here, an ideal flat beam is considered.

  c. Determine, as a function of V, the extraction voltage V_a for which $I_{1\,\text{max}}$ is obtained.

  d. Determine the beam width at $l_1 = Y_1/\beta_1$ for $I_1 = I_{1\,\text{max}}$.

Bibliography

V. E. Cosslett, "Introduction to Electron Optics: The Production, Propagation and Focusing of Electron Beams," 2nd ed. Oxford Univ. Press, London and New York, 1950.

A. B. El-Kareh and J. C. El-Kareh, "Electron Beams, Lenses, and Optics," Vols. I and II. Academic Press, New York, 1970.

H. Ewald and H. Hintenberger, "Methoden und Anwendungen der Massenspecktroskopie." Verlag Chemie, GmbH, Weinheim/Bergstrasie, 1953.

W. Glaser, Elektronen und ionenoptik, *in* "Handbuch der Physik" (S. Flügge, ed.), Vol. 33. Springer-Verlag, Berlin and New York, 1956.

J. J. Livingood, "Principles of Cyclic Particle Accelerators." Van Nostrand-Reinhold, Princeton, New Jersey, 1961.

J. R. Pierce, "Theory and Design of Electron Beams," 2nd ed. Van Nostrand-Reinhold, Princeton, New Jersey, 1954.

E. Segré, "Experimental Nuclear Physics," Vol. I. Wiley, New York, 1953.

A Septier, ed., "Focusing of Charged Particles," Vols. I and II. Academic Press, New York, 1967.

K. Siegbahn, ed., "Alpha-, Beta-, and Gamma-Ray Spectroscopy," Vol. I. North-Holland Publ., Amsterdam, 1965.

M. von Ardenne, "Tabellen zur Angewandten Physik," Vol. I. Deutscher Verlag der Wissenschaften, Berlin, 1962.

V. K. Zworykin, G. A. Morton, E. G. Ramberg, J. Hillier, and A. W. Vance, Electron Optics and the Electron Microscope. Wiley, New York, 1945.

Index

A

Aberration, 22, 23–26, 28, 30, 53, 92, 117
  aperture defect, spherical aberration, 25–26
  astigmatism, 25, 70
  chromatic, 26
  coma, 25
  distortion, 24
Acceleration
  axial motion, 18
  axis energy (potential), 4, 15, 44, 55, 63, 115
  preacceleration, 86
  system, 58
Accelerator focus
  acceleration–deceleration principle, 126
  source emittance, 17–18, 119–120
  space charge, 110–116, 119–120
Acceptance, 12, 32, 81–88, 91
Ampère's law, 43, 51
Analyzer concepts, general relations, 26–36, 77–88
Analyzer of sector type, 77–86
Analyzers
  electrostatic spectrometers
    coaxial cylinder, 96
    cylindrical sector, 73
    homogeneous field, 94
    spherical, 90
  magnetic spectrographs
    Browne and Buchner, 92
    with second-order focusing, 93
    semicircle, 33
  magnetic spectrometers
    lens, 101
    orange, 102
    semicircle, 29
  parabola spectrograph, 41, 94
  sector magnet
    inclined boundaries, 93, 103
    $n = \frac{1}{2}$, 89
    simple, 89
  velocity filter, 41
Apparent emittance, 22
Aperture
  defect, 25
  lens, 56, 58, 123
Astigmatism, 25, 70, 106
Auger spectroscopy, 118
Axial
  motion, 18
  symmetry, 21, 24, 44

144   INDEX

Axis energy (potential), 4, 15, 44, 55, 63, 115

B

Barber's rule, 90
Beam
　area, 84
　brightness, 12, 17, 131
　collimation, 12, 18, 86–88
　convergence (divergence), 8, 111, 125
　crossover, 85, 106, 111, 115, 116
　current (current density), 112, 116, 120–123, 125, 127, 130
　emittance, 11–22, 81–88, 119–120
　pencil, 17, 131
　perveance, 112, 125–126, 128, 130
　ray equation, 55, 58, 67, 115
　round (flat), 111
　scaling, 35, 112
　shape, 9, 64, 71, 113
　space charge, 18, 21, 110–131
　various effects, 117–119
Brightness, 12, 17, 131
Browne and Buchner spectrograph, 92

C

Calutron, 119
Cartan's construction, 105
Cathode, 17, 120
Central group of particles, 3, 11, 27, 40
Central path, 3, 27
Channel width, 27
Charge
　changing collisions, 117
　dispersion, 4, 41
Chromatic aberration, 26
Coaxial cylinder analyzer, 96
Collimation, 12, 18, 86–88
Coma, 25
Combined system, 5
Conservation of emittance, 16, 20–22
Convergence (divergence), 8, 111, 125
Current (current density), 112, 116, 120–123, 125, 127, 130
Cyclotron, 108

Cylindrical diode, 122
Cylindrical lens, 62
Cylindrical sector analyzer, 73

D

Deflection formulas, 38
Determinant, 6, 8, 13, 16
Diodes, 121–123
Dispersion, 4, 27–28, 40–41, 80
Distortion, 24
Drift region beam (round, flat), 110–120

E

Effective field boundary, 48
Electron, 17, 38, 112, 118, 120–127
Electrostatic analyzers, *see* Analyzers
Electrostatic field, *see* Field
Electrostatic lenses, *see* Lenses
Electrostatic quadrupole, 54, 68
Emission, 17, 120–121, 129–130
Emittance, 11–22, 32, 81–88, 119–120
　apparent, 22
　conservation, 16, 20–22
　determinant, 6, 8, 13, 16
　drift region (lens, electric double layer), 13–15
　focus width, 17–18, 119–120
　geometrical, normalized, 12, 16
　Liouville's theorem, 19–21
　one- and two-dimensional, 12, 20
　phase space, 18–21
　　axial motion, 18–19
　　transverse motion, 20–21
Energy dispersion, 4, 5, 27, 40, 80
Extraction, 120–131
　aperture, 123
　electron, 120–127
　ion, 127–131
　perveance, 112, 125–126, 130
　Pierce geometry, 124
　voltage, 126

F

Facet lenses (grid), 57
Faraday cage, 118

Field, 39–40, 42–54, 96, 111, 116, 121
  diode, 121
  drawing rule, 47
  equations, 42–43
  fringing, 47–51
  paraxial (plane symmetry, axial symmetry), 44–45
  Pierce geometry, 124
  quadrupole, sextupole, 51–54
  sector, 45–46
  space charge, 21, 111, 116, 121
  two-dimensional, 46–54
Focal surface (focal curve), 27, 33, 92, 95
Focusing
  acceleration–deceleration principle, 126
  drift region beam, 110–116, 119–120
  fringing field, 72, 103–109
  imaging (first-order, one-projection), 6–11, 29, 80, 90, 104–107
  intermediate focus, 116, 117
  lenses, *see* Lenses
  line focusing (point focusing), 22, 68, 70, 78–79, 105–107
  second-order, 24, 92, 93, 96, 100
  source emittance, 17–18, 119–120
  space-charge, 110–116, 119–120
  strong focusing, 109
Fringing field
  effective boundary, 48
  focusing, 72, 103–109

G

Gaussian image, 22
Gauss's law, 43, 44, 46, 111, 121
Goodness, 33–34, 85–88
Grid, 57, 88, 90, 96, 123, 124
Group of particles, 3–4, 11, 18–19, 40–41
Gun, 127

H

Homogeneous field
  electrostatic analyzer, 94
  magnet, 39, 49, 89, 92–94, 103–109
H-type magnet, 49

I

Image, 7, 22, 29, 30–31
  aberrations, 22–26
  formation, 6–11, 22
  Gaussian, 22
  imaging matrix, 5, 6–10
  resolving power (transmission), 26
  space (source space), 6
  stigmatic, 22
Immersion lens, 62
Inclined field boundaries, 50, 72, 93, 103–109
Ion, 38–39, 112, 117–119, 127–131
Isotope separator, 94, 117, 119, 128

L

Laplace equation, 43, 124
Lens(es), 55–72
  acceleration system, 58
  aperture, 56, 58, 123
  cylindrical, 62
  grid facets, 57
  immersion, 62
  inclined field boundary, 72
  magnetic, 65
  quadrupole (singlet, doublet, triplet), 67
  thin lens approximation, 56
  unipotential, 62
  weak lens, 58, 67
Line shape, 29–32, 87
Liouville's theorem, 19–21
Luminosity, 12

M

Magnet, 38–41, 49, 89–94, 103–109
Magnetic field, *see* Field
Magnetic lens, 65
Magnetic quadrupole, 52, 67–71
Magnetic spectrographs (spectrometers), *see* Analyzers
Mass dispersion, 4, 41, 88
Mass to charge ratio, 37, 39
Matrices, 3–10

Motion
  axial, 18–19
  nonrelativistic, 35–37
  relativistic, 37–39, 108
  thermal, 17–18, 37, 120, 121
  transverse, 20–21

N

Neutralization, 117
Nonuniform current density, 117
Normalized emittance, 16, 20
Normalized perveance, 112

O

Optics (first-order, one-projection), 6–11, 16
Orange spectrometer, 102

P

Parabola spectrograph, 41, 94
Paraxial fields, 44–46
Paraxial ray, 4, 22
Particle
  groups, 3–4, 40
  relativistic, 37–39, 108
  scaling, 35, 112
Pencil beam, 17, 131
Perveance, 112, 119–120, 125–126, 130
Phase space, 18–21, 81–84
Pierce geometry, 124
Plasma ion source, 127–131
Preacceleration (preretardation), 86–88
Projection, 3, 5, 12, 16, 20

Q

Quadrupole
  field, 52, 54
  lens (singlet, doublet, triplet), 67–71

R

Ray
  equation, 55, 67, 112, 115
  paraxial, 4, 22
  virtual (source space, image space), 6

Relativistic velocities, 37–39, 108
Residual gas, 117
Resolving power
  imaging, 26
  spectral, 29–32, 80, 85–88
Retardation, 115, *see also* Acceleration
Richardson's equation, 121

S

Saturation emission, 120, 121
Scaling, 35, 112
Scattering, 117
Second-order focusing, 24, 92, 93, 96, 100
Sector
  analyzer, 73, 77, 89, 103
  field, 45–46, 50
  focusing, 108
  magnet, 89, 103
Semicircle spectrometer (spectrograph), 28, 33
Sextupole, 52
Similarity, 35–36
Slit
  aperture lens, 58
  source, 119
Source
  emission, 17, 120, 130
  emittance, 17–18, 119–120
  space (image space), 6
Space charge, 18, 21, 110–131
  compensation, 118
  conservation of emittance, 18, 21
  diodes, 121–123
  drift region beam (round, flat), 110–116
  electron extraction, 123–127
  ion extraction, 127–131
  low-emittance beam, 119–120, 126
  nonuniform current density, 117
  perveance, 112, 119–120, 125–126, 130
  Pierce geometry, 124
  scaling, 112
Spectrographs (spectrometers), *see* Analyzers
Spherical aberration, 26
Spherical analyzer, 90

Spherical diode, 122
Stigmatic focusing, 22, 70, 78–79, 105–107
Strong focusing, 109
Subsystem, 5–6

T

Tandem accelerator, 117
Terminal voltage, 40
Thermal motion (thermionic emission), 17–18, 37, 120, 121
Transmission
  imaging, 26
  particle analysis, 32, 85–88
Triode, 123

U

Unipotential lens, 65
Units, 38, 112

V

Velocity
  filter, 41
  relativistic, 37–39, 108

W

Weak lens, 58, 67

X

X-ray production, 18, 118

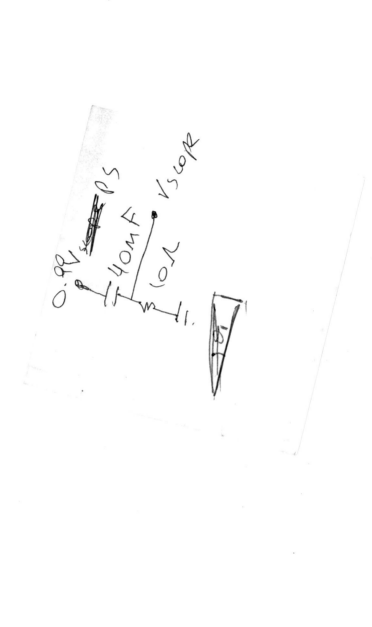

233645